AF690794

Marketing Rewired

Why you need to think differently in the Age of AI-Driven Marketing

Published by: CMFG Management Ltd.

Marketing Rewired

Why you need to think differently in the Age of AI-Driven Marketing

For permission requests or commercial licensing, please contact:greg@cmfg.co.uk

978-1-0682301-4-1

Foreword

I've spent my career observing how marketing has evolved, but seldom has the change seemed so urgent. Buyer behaviour, technology and societal expectations are all changing in ways that make old playbooks redundant virtually overnight.

What you have here in your hands is more than just a marketing guide. It's a challenge to think differently, not in the cheap way of chasing trends but in the meaningful way of asking new questions, reframing problems, and seeing principles over tactics.

This book accomplishes what few others do: it balances crisp strategic thinking with on-the-ground practicality. It uses data, psychology, technology and lived experience to offer a clear-sighted path forward whether you're trying to build a brand, get your team on the same page or redefine your role in the business. I have focussed mostly on how things apply to B2B marketing, but it applies just as much to marketing in general. Marketing as a whole is changing rapidly, don't get caught in the blocks.

In a world ever more ruled by machines, the future of marketing still belongs to those who can think. This book will help you in that mission.

Dedicated to my wife Melanie,
my son Alexander and
my daughter Ellen.

Contents

Introduction: B2B Marketing is Undergoing a Seismic Shift

"Don't let yourself make excuses for not doing the things you want to do."

Sam Altman, CEO of OpenAi.

The complexity of modern marketing, coupled with the emergence of Artificial Intelligence (AI), calls for a radical rethink of established practices.

AI hasn't just added another tool to the marketer's belt, and it's rewiring how we think about brand, buyer behaviour, customer experience, and strategic planning. But the real transformation isn't about the tech. It's about mindset.

Old frameworks built around linear funnels, quarterly campaigns, and hard-sell tactics are breaking down. In their place, a new marketing model is emerging, one powered by intelligence, designed for trust, and built to flex in real-time. The use of AI in marketing has exploded and is having impact across the board. https://digitalmarketinginstitute.com/blog/10-eye-opening-ai-marketing-stats-in-2025

Whether you're leading a global team or scaling a startup, your job as a B2B marketer is no longer just to broadcast or convert. It's to anticipate, engage, personalise, and adapt, often in milliseconds.

Marketing Rewired is not a how-to manual. It's a strategic guide for those who want to lead through change. You'll find a modern approach that integrates behavioural science, brand strategy, AI enablement, and real-world application. It's grounded in commercial thinking, informed by research, and built for the complexity of today's B2B buyer.

Inside, you'll learn how to:

- Move beyond vanity metrics and short-term campaigns

- Understand the hidden forces driving buyer decisions
- Align sales, marketing, and success around shared revenue impact
- Use AI not as a replacement but as a force multiplier
- Build long-term brand trust while still accelerating growth

But perhaps most importantly, you'll be asked to reconsider your assumptions about creativity, automation, data, and what it means to lead marketing in this new era. I have worked in the industry for many years and have seen it change many times, but not like this and at this pace.

I have used many links to illustrate points, back up assumptions, show you surveys, guides, case studies and reports. All the links are live at present, but they can change. If they do the text of the link will give you an idea of what to look for.

If you're ready to move beyond best practice and start shaping the next practice, this book is for you because, in the era of B2B, best practice won't be good enough.

Let's begin.

1. Think Differently About B2B Marketing

"Marketing is no longer about the stuff that you make, but about the stories you tell."

Seth Godin, Marketing Guru.

The Rise of the Strategic Mindset

The expression "think differently" is often wielded as a rallying cry for creativity, but in the race to remove friction from the buying experience, it's a necessity for B2B marketing. The radical changes in buyer behaviour, technology, global business expectations and the progression of AI has changed the playbook. When I started in the industry, the traditional methods in place were using linear funnels and standardised campaigns trying to pursue larger volume-based KPIs that are no longer showing steady returns. In their place, a new era of marketing has arrived, one that demands an approach based on adaptability, foresight and integrated intelligence.

B2B marketers in today's world work in a fluid and ever-evolving environment where the lines between sales, marketing, product and customer experience are blurring more than ever. To win, you need a new strategic lens, one that not only pulsates with digital technology but also uses it to envision a new way of delivering value across your customer life cycle. Marketers need to view themselves as growth orchestrator's, not just communicators/campaign managers.

Thinking differently starts by rethinking marketing itself. For far too long, marketing has been limited to brand awareness, lead gen, or sales enablement. However, in a world-class organisation, marketing is transformed into a business-strategy driver integrated into strategic planning, revenue operations, customer experience, and innovation. This shift means marketing is no longer seen as a cost centre but directly contributing to value creation and even driving direction at the board level.

AI is central to this strategic redefinition. From chatbots and predictive email subject lines, AI has evolved to provide deep customer insights, smart segmenting and personalisation at scale, and real-time content cues. A recent analysis by **McKinsey** showed that companies that use AI in marketing and sales create 50 percent more leads and appointments, while the cost associated with creative production can fall by up to 60 percent. https://www.mckinsey.com/capabilities/quantumblack/our-insights/global-ai-survey-ai-proves-its-worth-but-few-scale-impact

These are not simply gains in productivity; they also reveal a shift in how customer relationships are built and maintained.

But AI is not a plug-and-play solution. It needs to be part of a broader strategic fabric that takes into account human insight, brand clarity and business alignment. Marketers need to find a way to couple automation with empathy. Yes, data can predict behaviour, but it's still up to a marketer to interpret intent, shape messages that resonate, and exercise judgment on what's right and wrong, when and how to engage.

This is where the concept of augmented marketing leadership comes into play. Today's B2B marketing leaders must be comfortable wearing many hats. They need to be fluent in data science, agile ways of working, behavioural economics, sustainability, brand architecture, and commercial performance. Thinking differently is about building that range, not as a catalogue of technical skills, but as a leadership capability that leads to cross-functional unity and long-term competitiveness.

This change of mindset needs new culture in practice. Such as shifting from risk-averse to experimentation-led and accepting that not everything will pay out an immediate ROI. Ask any CMO why they renewed a contract, service or platform and they will rarely mention anxiety. But that's often the real reason – what if it goes wrong and I'm to blame? Next-generation companies empower marketing teams to prototype new platforms, try new disruptive messaging, and co-innovate with customers. The process of thinking differently is to create a learning organisation that learns at the same pace as the rapidly changing marketplace.

Additionally, businesses need to work against the jargon they employ internally. Instead of tactical sound bites, they should start using the language of strategy, risk, and revenue. By demonstrating the impact that marketing investments have in growing the business, retaining customers, increasing share of wallet, and lifetime customer value, marketing can seek the seat at the table that is due. That is especially important because AI transformation is something that touches every part of the business.

At its core, doing new things in B2B marketing does not mean reacting to disruption. It is active about the future. It is about imagining what marketing will look like if machines enhance decision-making, buyers act as networks, and value sits across ecosystems and is co-created. It's about transforming marketing into an operating system of intelligence, where every level of the business is monitoring buyer signals and being always informed by market trends and empowering rapid response.

Through the pages of this chapter, we'll show how that mentality applies to the marketing discipline, from buyer psychology and content strategy to sales alignment and brand positioning. Yet here's where it starts to change, with a strategic migration: to realise that today's marketer is not someone who builds campaigns but rather leads change, thinks in systems and organises a culture.

The Buyer Has Changed

Much has been written about the changing nature of B2B buyers, but many organisations still aren't changing their marketing approach to reach this new reality. The traditional funnel, the linear path from awareness to consideration to purchase, implies a level of linearity and control that isn't there anymore. Today's buyers are empowered, autonomous, and more discerning with how or with whom they choose to interact.

Gartner reports the average B2B buying journey involves 6.8 stakeholders, each armed with four or five separate pieces of information they've gathered independently before even connecting with a supplier.
https://advertisingweek.com/the-state-of-the-enterprise-b2b-buyers-journey/

Such an influence network renders decision-making much less predictable and more intricate than legacy models can handle. According to **Gartner**, customers spend just 17% of their time meeting with suppliers combined in the lead up to a purchase and 75% of B2B buyers prefer a 'rep-free' sales experience.
https://www.gartner.com/en/sales/insights/b2b-buying-journey

The people who buy today are more like researchers or curators than receivers of information. They expect to self-educate compare, verify and engage on their own terms. This transformation means that marketers must abandon trying to push messaging and instead focus on how to create environments where trust can be nurtured over time. Content has to be findable, relevant, and deep, allowing the buyer to consume at their own speed and their specific level of interest.

Yet despite this, there are still many B2B companies that gate their best content, frustrate users with clunky forms, or design journeys based on how their company is structured rather than on customer intent. A new way of thinking implies a dramatic new way of looking at things: Marketers need to innovate from outside in. This involves putting the customer experience ahead of the convenience of the department and facilitating an exchange of value which recognises that buyers are giving you their time and attention.

It also means that brand matters more than ever. When the world is a noisy, broken place and products and services are easily commoditised, buyers use trust, reputation and emotion as your decision-making filters. B2B brands have long been guilty of underinvestment in emotional connection, believing that they can get by through communicating in a rational way. But behavioural science, as well as empirical evidence, increasingly suggest that emotional brand associations are absolutely critical to the complex decisions people make, even those made by senior executives and technical audiences.

According to the B2B Institute, a joint effort with Ehrenberg-Bass Institute, the 95:5 Rule illustrates the concept that at any point in time, just 5% of companies in your target market are in any state to consider buying products or solutions like yours, and 95% are not. So, a successful marketing strategy does not focus only on those buyers who happen to buy immediately but also builds brand awareness and trust among future potential customers. https://business.linkedin.com/marketing-solutions/b2b-institute/how-b2b-brands-grow

This realisation reiterates the need for long-term brand building as well as short-term sales tools in B2B marketing strategies.

AI, again, has an enormous advantage. Predictive intent data tools, like **Bombora** or **6sense**, behavioural analytics solutions such as **Hotjar** and **Heap**, and customer data platforms (CDPs) like **Segment** can reveal hidden signals, enabling marketers to trigger early engagement, inform intelligent personalisation and help accelerate time to value. But these tools also need to be wielded responsibly. Buyers are increasingly sensitive to how their data is used, and overreaching, or a lack of transparency about how data is handled, can lose trust quickly.

Personalisation should, therefore, be based on relevance rather than mere data points. This means understanding not only the company or title, but the decision context: What's this person's role in the buying group? What challenges are they confronting in their industry? What are their internal stories telling them about what is important? These levels of personalisation require collaboration between marketing, sales, customer success and product teams, and a shared dedication to stop approaching buyers as a persona but instead see them as people.

One way to do this is by creating content journeys that mimic real behaviour as closely as possible. That encompasses ungated resources that create authority and goodwill, interactive tools for potential buyers to assess fit, and peer-driven content such as customer reviews, video testimonials, and community insights. These assets don't just entertain, they earn trust.
Thinking differently about buyers also requires understanding the entire

customer life cycle. Far too often, marketing is organised around acquisition, passing along leads and then retreating post-sale. But today's marketing is not just about the deal. Retention, onboarding, expansion, and advocacy are all so important to sustainable growth. In reality, a great deal of the most successful marketing strategies today are based on customer success - ensuring that the post-sale is so awesome that it provides fuel for referrals, renewals and net new business.

For this to be possible, B2B marketers also have to use data and AI to keep connected with customers even after they sign the contract. Are users embracing your product? Are they seeing value? Are they satisfied? Are they engaged? Technologies such as **Gainsight** and **Totango** provide the ability to monitor customer health and trigger marketing actions in response, from education campaigns through incentive offers to invitations to participate in case studies.

The bottom line is this: the millennial B2B buyer is not evolving, they have evolved. The question is whether your marketing has. It involves not just challenging what we've always thought but abandoning the worn-out assumptions, investing in the currency of relevance and trust, and designing every touchpoint around how buyers act with us rather than how we wish they would.

Creativity and Courage in the Age of Automation

As B2B marketers begin to put AI, automation and data-driven decision-making to good use, there is a danger that we will lose that component, which is so fundamentally human that it ensures a brand resonates and sticks in our minds all the better for this emotional connectivity. Thinking differently, in this case, does not mean getting rid of technology but redefining its role and incorporating it into creativity. Indeed, the necessity for imaginative thought in B2B has never been more vital.

In a world of content saturation and algorithmically refined messaging, what dominates is not necessarily the most efficient message but rather the most surprising one. The one thing is creativity. It gets noticed, remembered and felt. However, creativity has traditionally been treated with caution in B2B marketing. Messaging can be too safe, too buttoned-up and too technically minded, often on the premise that B2B buyers make purely rational decisions.

The justification is increasingly challenged by evidence. According to a **Google** and CEB report, B2B buyers tend to exhibit a higher degree of stress and/or pressure when purchasing items than B2C consumers, as there's often more personal career risk and long-term impact with a business purchase. Emotion is not only part of B2B, but it may also be even more important than in consumer markets.

Creativity, they say, is not a soft skill. It is a strategic tool. It offers a filter for marketers to analyse intricate customer insights, generate unique messaging and connect with audiences through authenticity and empathy. Even beyond that, it creates brand capital itself, which is a well-established driver of pricing power, customer loyalty, and share of voice.

Campaigns that dare to be different have shown remarkable results in B2B. Then there's **Slack**'s "So Yeah, We Tried Slack" campaign, which managed to stand out among the crowded category of collaboration tools by utilising humour, social proof, and storytelling. Or **Adobe**'s "The Drumbeat of Business," where visual metaphor and rhythm were employed to dramatise the power of data-driven creativity. No, these weren't "quirky for quirky sake" – they were based on audience insight, came in an original way and then were turbocharged by technology.

But creativity is not enough on its own. The other half of the equation is bravery. None of those things are easy. It's brave to go against the grain, to pitch a risky idea to stakeholders, or to run a brand-led campaign when everyone else is shouting about features. In a field where marketers are routinely evaluated based on short-term performance measures, bravery might seem like a risk. But it is also what distinguishes between brands that lead and those that follow.

This is where artificial intelligence can surprisingly come in to help. AI, when used thoughtfully, can free up time for creativity. It can systematise some of the low-value work, analyse those performance trends, and begin to produce first draft content so that marketers have more time to perfect creative, experiment with new formats, and double down on brand stories. Tools like **Jasper** and **Canva**'s Magic Studio might help support rapid prototyping, and platforms like **Persado** can use natural language generation to test how messaging resonates emotionally.

However, creativity should never be completely outsourced to AI. "AI in creative workflows' most effective place is as a copilot, not creator. It's human marketers who need to drive the emotional arc, fashion the strategic narrative, and infuse their brand voice with authenticity. Without that human side, the most technically proficient of campaigns will not resonate.

Thinking differently also builds a culture that nourishes creativity. Far too often, creativity is confined to "the creative team" or farmed out to agencies. In today's B2B organisations, creativity is everyone's job. Product managers, analysts, and sales reps, can all help with the insight, ideation and messaging building.

And success in creative in B2B is often about internal buy-in, too. That's why marketers need to be not just creative minds but also creative diplomats, able to push for brave ideas with business rationale. Today's most successful CMOs know how to talk the CFO talk while advocating for creative ambition. They create business cases for brand investment, take customer evidence into strategy meetings, and argue about the power of originality in the boardroom.

It is also important to remember that creative courage is not about spending millions of dollars. Some of the best campaigns come from the tightest constraints. One line of well-placed copy, an explosive **LinkedIn** post, and an amusing use of data can make a powerful impact. It's not always the size that matters, but substance.

Creativity is one of the few things about a company that a competitor cannot reproduce cheaply and that a company cannot redistribute throughout other business units. Technology can be bought. It is possible to minimise distribution. But creativity, real creativity, emotionally intelligent, culturally attuned creativity, is a long-term differentiator. It humanises the brand. It builds advocates. It earns attention and trust.

As AI matures, the pressure will mount to automate a larger percentage of the marketing mix. But to think differently is to resist being drawn toward sameness. It means doubling down on what makes your brand unique. In this new world, creativity is not simply a luxury. It's a commercial necessity.

From Lead Generation to Commercial Impact

That is one of the defining shifts in B2B marketing right now; in moving from lead generation to this bigger, more strategic cross-functional commercial growth muscle. In a lot of companies today, marketing is still measured by how many leads it can create or the traffic it is able to bring in. Although these metrics are simple to record, they are not always predictors of success in the long run.

Thinking differently is leaving vanity metrics and siloed objectives behind. It means repositioning marketing as a commercial partner together with sales around revenue, profitability, and lifetime customer value. This isn't theoretical; it is already taking place in high-performing organisations.

The development of revenue marketing and Revenue Operations (RevOps) has driven this change. These models bridge the gap between sales and marketing, customer success, and even product. They consolidate data, operations, and reporting for a common commercial purpose. This places marketing as a co-owner in pipeline generation, deal velocity, churn reduction, and customer expansion.

This is where the magical combination of AI and automation emerges. No longer are they lead-scoring tools, now they are revenue intelligence engines. Platforms like **Salesforce Einstein**, **HubSpot**'s Operations Hub and **Gong** consolidate the behavioural signals across the customer journey, providing insight into which accounts are going to convert, renew or churn. These observations may inform how marketing teams move from reactive campaign delivery to proactive growth planning.

But technology won't change the outcomes on its own. It doesn't just require a behaviour change, one that embraces financial fluency. They need to get used to talking about terms like unit economics, CAC (customer acquisition cost), gross margin contribution, and ROMI (return on marketing investment). These aren't simply finance terms; they are crucial statistics you'll need to input into your strategic marketing planning.

For instance, if a product experiences high churn or low renewal rates, dumping budget into top-of-funnel awareness could be a waste. Instead, marketers could shift attention to onboarding campaigns, educating customers, or upsell based expansion tactics. Similarly, a low-penetration, high-value segment could also justify a custom story, sales enablement and focused thought leadership pieces. To think more commercial is to prioritise your marketing work based on business impact rather than just channel efficiency.

It also means rethinking the antiquated handoff process between marketing and sales. The lead handoff is still a hard stop in a lot of organisations – an arbitrary line in the sand beyond which the responsibility is sales rather than marketing. But buyers don't live trips that way. They fluidly cross channels, recall old content, and involve as many dozen chefs as possible across departments. Today's marketing team needs to operate in constant partnership with sales around pipeline progression.

Not only does this require us to align our processes, but we also share responsibility. As a revenue team, we will have sales engineers, growth specialists, and customer advocates who work to move the needles that matter. The work of a traditional sales team is not enough, which is why revenue-focused teams can no longer afford to go it alone. In these spaces, marketers create content, insights and digital experiences that advance deals. Thinking differently also means challenging our ways of measuring success.

Classic dashboards still favour metrics like cost per lead or MQLs, which might overlook quality, velocity or downstream conversion of opportunities. Most mature companies will monitor things like pipeline coverage (amount of pipeline vs target), conversion rates by campaign, average deal size by persona, marketing influenced revenue, etc. They are proof points which make the complete commercial contribution of marketing over the life cycle tangible.

Sustainable marketing infrastructure is also important for long-term success. That investment includes investments in first-party data, content management systems, journey orchestration platforms and customer data platforms (CDPs). Such tools give marketers a way to own their audiences as third-party cookies disappear and privacy laws take hold. They also facilitate more granular segmentation, more contextualised personalisation, and better performance measurement.

Furthermore, marketers need to be more integrated in pricing, product positioning and marketing strategy. The commercial impact isn't just about converting interest; it's about creating demand, teaching buyers and swaying perceptions of what's valuable. In collaboration with product and commercial leaders, marketers can develop messaging frameworks, uncover new market opportunities, and help tune sales approaches towards value.

Finally, these organisations want the marketing function to be responsible not just for growth, but for sustainable, profitable growth. That requires achieving a balance between acquisition and retention, brand equity and performance marketing, and long-term investments and short-term wins. It also involves creating an internal culture of experimentation, agility, and accountability.

In other words, the days of the lead factory marketer are gone. The future is for marketers who are able to see the full commercial picture, who can sit in the boardroom, affect investment, and deliver results that are many steps beyond awareness. Thinking differently means taking responsibility for the business impact, not just brand metrics. And it means walking into the boardroom with confidence, not just into the studio with creativity.

Thinking Differently Is a Discipline, Not a Moment

"Think different" is easy to say. That sounds like something bold and new, something exciting. But in reality, it is not so much a mindset as a discipline in B2B marketing that has meant operationalising curiosity, embedding agility and institutionalising long-term thinking. It means always showing up prepared to question assumptions, adjust strategies and lead with integrity, even when the answers are uncertain.

All too frequently, a different way of thinking is interpreted as chasing the latest great idea. Marketers jump on the next new platform, tool, or buzzword to try to leap ahead. 'Growth Hack' is my personal hated term. But authentic differentiation isn't a reaction. It's from consideration and purpose. It stands on principles, not passions. To think differently, marketers not only have to apprehend a changing environment but also be prepared to separate from it when it doesn't suit.

This is a culture of learning. The top marketing teams of today aren't the ones with the biggest budgets or the most impressive campaigns; they are the teams that learn quickly. They try and test and try and test and analyse and adapt. They create psychological safety so that people are game to take risks. They support cross-functional learning, customer co-creation, and strategic retrospectives.

Thinking differently also entails embracing complexity. Truth be told, the B2B world is anything but black or white. Considerations are clouded by various priorities, legacy platforms, geographic homogeneity or heterogeneity and buyer mindsets. The best marketers don't try to simplify complexity by obfuscating it. They face it with empathy, rigour and pragmatism. But that's not the same as paralysing acceptance. Exceeding things in the name of marketing becomes a by-product of over-engineering: long strategy decks, months of planning, and over-analysis of attribution data. Speed matters. So does momentum. Thinking differently is about striking the right balance between strategic depth and creative speed, between reflection and action.

It also requires courage. Change is seldom an accident in complex systems. It must be advocated for. This means challenging archaic processes, demanding better customer experiences, and holding the line on long-term brand investment even when the quarterly numbers call for quick wins. Courage is not loud. It is often quiet. It is in deciding what is most important and in defending it when others look elsewhere.

Today's B2B CMO is no longer just a communicator but a change maker. They have to break down silos, unify sales, product and customer teams and help the business navigate complexity and change. They need to exemplify this type of team-based, strategic thinking that takes marketing out of the t-shirt-branded boardroom.

AI is driving the necessity of this change ever faster. Thinking differently is necessary when decisions are increasingly made by algorithms. Now marketers must wonder: Are we even feeding AI the data it needs? Are we constructing ethical guardrails? Are we second-guessing the outputs of AI with human judgment? In the AI era, thinking differently means making sure that machines serve strategy, not the other way around.

It also means staying human. For example, the most advanced AI in the world can't mimic authenticity, creativity or empathy. These are the marketer's perennial superpowers. To think differently is to consider employing AI to amplify, not replace, the human elements that make marketing matter. This means being able to communicate in ways that touch us, being able to build brands we care about, and being able to win trust in a world where we have so little time to pay attention and so much mistrust.

Ultimately, the acceptance to think differently is not a one-time idea or even a call to action; it is a professional commitment. A way of working. An unwillingness to settle for average. It's what allows B2B marketers to move from checkpoint to outcome maker, from number-reporter to strategic influencer, and from content monkey to movement maker.

Summary:

- Old B2B marketing systems are broken.
- Thinking differently isn't a nice to have; it's a strategic imperative in an AI-enabled, buyer-centric world.
- AI is incredibly powerful when it comes to personalisation, insight, and automation, but we must not forget to counter-check human strategy and ethics when applying these tools.
- Creativity has become a commercial differentiator, not a luxury. It's the sort of brave, emotionally compelling marketing that cuts through noise and drives long-term brand equity.

Case Studies in Action:

Adobe launched the "Creativity for All" campaign on TikTok to celebrate and encourage creativity among its users. The campaign featured challenges and tutorials using Adobe products, spotlighting creators from diverse backgrounds and skill levels. It leveraged user-generated content to show the possibilities of Adobe's software suite. https://camphouse.io/blog/best-marketing-campaigns

Toucan, a French analytics software provider, integrated **Cognism**'s AI-powered sales intelligence to enhance lead generation and conversion rates. By using **Cognism**'s platform to identify high-intent accounts and prioritise outreach, Toucan saw a 30–50% increase in its leads-to-meetings-booked rate. Additionally, 20% of the company's revenue was directly attributed to campaigns powered by **Cognism**, and the team reduced their sales technology overhead by 40% through consolidation.
https://www.cognism.com/blog/toucan-case-study

2: Inside the B2B Mind: Psychology, Bias and Behaviour

"Instead of interrupting, work on attracting."

David Meerman Scott, Marketing Strategist.

Behind every successful marketing strategy is an accurate perception of how B2B customers buy. But paradoxically, for all those assertions about how B2B buyers are rational and data-driven, the psychology of B2B purchase is a messy thing that is hard to nail down. Whereas a B2C consumer makes an emotion or impulse decision, B2B buyers run through a minefield of internal approvals, risks, group-speak and rubber stamping – all while adding a massive layer of psychological complexity to the process.

You are not the only one making the buying decision. B2B purchase decisions are rarely made by one person. Instead, it's one of many negotiations shaped by a raft of people from across the various departments, all bringing their own perspectives, priorities and biases to the discussion. The finance will look at the cost line, IT will evaluate the technical fit, procurement will work for attractive contract terms, and the business will look at the long-term return on investment. The result is a slow, complex, drawn-out process that can last for months or even years, depending on how complex the purchase is and how much investment is required to finalise it.

While decisions to buy and sell B2C may be influenced by a whim, based on convenience or simply liking someone, B2B buying is strategic. All changes must be aligned with the Strategic Business Objectives, operational needs, and long-term organisational goals. Because many B2B deals are large in ticket size (monetary value), the stakes are high. A poor decision can lose you money and time or tarnish your reputation, risks few in the business world want to take.

What this means is that the management of risk and the construction of trust are as important as the features or lines of a product or service. Buyers are not just evaluating vendors based on what they have to offer; they are also assessing product reliability, industry reputation and the opportunity for a long-term, secure relationship. That's a big part of why certain vendors seem to thrive even when their offerings are as good (if not a little worse) than their competitors, and it all boils down to the psychology of B2B sales.

In this chapter, we take a deep dive into the psyche of the modern B2B buyer. We will explore the role of cognitive biases and the balance between rational and emotional, the factor of trust, and how organisations can leverage this information to create better marketing and sales. Knowing and recognising these psychological factors can help you as a B2B marketer create a more powerful position, reduce sales cycles, and build long-term loyalty.

B2B Buyer Decision-Making: The Psychology Behind the Process

B2B Purchasing and the Power of Cognitive Biases

B2B buyers are all human...and human-based cognitive biases are a big part of how they consider (and ultimately decide between) vendors. Such biases can affect our preferences, risk attitudes, and decisions, which are generally under the radar of our conscious awareness. As a B2B buyer, you're subject to a number of very well-understood cognitive biases which govern your decision-making process, these are:

Anchoring Bias

Buyers rely too heavily on the first piece of information they receive, which becomes a yardstick against which they measure all other information that comes their way. That central truth is that the way a product is priced or positioned in the beginning is so enormously influential.

Application: Marketers must lead with price and value premises early so that the perspective on the customer is set up.

Confirmation Bias

Leaders seek information that confirms their existing opinions and dismiss contrary evidence.

Implication: Marketers will need to create follow-up content, supported by data and customer testimonials, to dispute these old assumptions and to get people willing to take the next steps.

Loss Aversion

For more people, the fear of choosing wrongly outweighs the attraction of choosing rightly. Fear prevents people from changing vendors too often.

Application: Risk reduction tactics are shown to lower perceived risks (risk-free trials, guarantees, case studies) and increase the fulfilment of requests.

Social Proof & Herd Mentality

When making a choice, customers tend to follow their rivals and market leaders.

Application: Credibility is established by the most prominent industry players' testimonials, case studies, and customer quotes.

The Bandwagon Effect

How easy it is for people to trust and follow each other.

Use Case: Prospects may feel safer when choosing a solution if they know it's something that other industry leaders or competitors are using.

Status Quo Bias

B2B purchasing behaviour is generally characterised by resistance to change because changing out a supplier is both an effort and a perceived risk. In fact, so many of them don't end in a decision to buy at all. Globally, many B2B purchase processes end with 'no decision' made thanks to the 'stickiness' of the current state.

Application – By focusing on ease of change, migration support and integration are serious palliatives to break down this resistance on behalf of the marketers.

Emotive And Rational Reflection in B2B Purchase Decisions

While B2B purchase decisions appear on the surface to be rational decisions, more than anything else, buying decisions are emotional. Rational analysis and emotional drivers are what and when a company decides to commit to a purchase. The marketers that can bring both of these two dimensions into balance are far better equipped to produce compelling, successful campaigns - and also drive conversion and customer success.

Critical decision-makers have to ensure their decisions are cost-effective, operationally feasible and will further the company's goals. Five things to consider are:

- **Cost benefits and ROI:** Purchasers ask whether a product will provide measurable economic benefits with a reduced cost or an increase in revenue.
- **Details and specifications of the product:** IT managers and purchasing teams at the Company need to know whether the solution will fit well into their current set-up.
- **Consistency with industry norms:** Presumably, some firms face quite stringent industry norms, and the question is not really if in any but rather when.
- **Scalability and ongoing support:** Companies are looking for solutions that will grow with them, that the service or product is ergonomically friendly to the company as it grows and continues to support them into the future.
- **Procurement and contract review:** Legal and procurement squads are investigating whether the purchase fits what is covered in company policies and contract obligations.

These are important rational constraints, but it also doesn't seem to drive the final choices that people make. Rather, they are the threshold expectations under which suppliers need not apply. **EMOTIONS TRUMP EVERYTHING.** The technical guy is the waiter at the wedding.

The Emotional Thought Process in B2B Procurement

B2B purchasing might be done on an organisation's behalf, but the people making the decisions are still people. One such factor involves personal or emotional reasons, such as trust, reputation and the feeling of safety.

Among Emotional Drivers are:

- **Trust in supplier's brand & reputation:** Buyers want to buy from companies they know well and have a reputation.
- **Fear that if they make a mistake in their selection, they will suffer negative consequences:** Those who make supplier decisions often believe that making a bad supplier decision will make them look bad as well as damage their career.
- **Relationship confidence and supplier support:** Your purchases and their arrangements are likely to lead to retained relationships over time, so both customer service and how they manage the relationship will be important.
- **The alignment of a supplier with culture:** Suppliers whose values align with the culture of the buyer are also more likely to be taken to market and through an evaluation, particularly if the values of the supplier are around sustainability, ethical business or innovation.
- **Personal prejudice and in-house political pressures:** Decision makers can be influenced by previous experiences with suppliers or recommendations from colleagues.

More and more research indicates that gut feelings and instincts heavily influence B2B decision-makers rather than logic alone. In a lot of cases, you chose a supplier not because they are the best price or best feature set but rather because they are the company that you've come to trust and rely on. https://www.forrester.com/blogs/who-do-b2b-buyers-trust/

How to Cater for the Rational and Emotional Factors

To succeed in B2B marketing, businesses need to create strategies that combine the rational and emotional aspects of decision-making.

Here's how:

Rational Appeals:

- **Produce Thought-Leadership Content:** Share thought leadership content based on data and industry expertise, including white papers, case studies, and research reports.
- **Offer Clear ROI Metrics:** Financial modelling, case studies, and references will support showing the return on investment.
- **Be transparent:** Be clear on the cost, how it works, the terms and risks.
- **Provide product demos & trials for Technical Documentation & Trials:** Give prospects a chance to kick the tyres and test out the product to see if promises are truly being met.
- Simplify Procurement and Compliance: Provide compliance certifications and contract flexibility to help make it easier for your legal and procurement department to approve the purchase.

Emotional Leveraging:

- **Establish Brand Credibility:** Regularly build trust with awards, endorsements and press.
- **Harness the Power of Storytelling:** Relate actual customer success stories that capture the heart, such as triumph over adversity or meeting with great success.
- **Relationships:** Communicate 1:1, have executive involvement and make sure help is on its way to your customer proactively.
- **Experience, Adoption and Retention:** Deliver a strong customer service, onboarding and training program to mitigate buyer anxiety post-purchase.
- **Generate a Feeling of Community:** Use networking events, user groups, and premium customer forums to tie the buyers to your product community and loyalty to it.

The Importance of Psychological Safety in the B2B Sale

Psychological safety has a big impact on the decision-making process and buying. People need to be confident that the life-or-death decisions they make will not result in anything bad happening to them. Sellers who can distance themselves from uncertainty and gain buyer confidence have a competitive advantage.

Some of the tactics to build psychological safety are:

- **Offering Social Proof:** Some lifters are looking for proof of achievement, so show them case studies, testimonials, and/or endorsements from reputable industry leaders.
- **With a 'No-Risk' Guarantee:** Money-back guarantees or performance-based contracts can disarm fear and increase commitment.
- **Display Long-Term Stability:** When you demonstrate the financial stability of your company and its roots and lineage in the marketplace, potential buyers are much more credible and authentic.
- **Pilot Programs:** Enabling companies to try a solution on a limited scale before full deployment lowers perceptions of risk.

Combining rationalisations with emotional comfort allows companies to bypass decision inertia, alleviate buyer's remorse, and create long-term engagements. Companies that don't handle both sides lose prospective customers, regardless of whether their product is superior.

Trust & Brand Reputation

Trust is a major consideration in B2B purchasing decisions. In contrast to B2C sales, where trust can be established via reviews and personal referrals, B2B deals can be long-term commitments and high investments with much at stake. Buyers, therefore, want to be sure of the vendor they choose before making a purchase.

Trust is not only the basis for business relationships but drives several aspects of the purchase process:

- **Lowered Perceived Risk:** Purchasers tend to feel less at risk when they select a vendor with a proven track record and are less worried about losing money or project failures.
- **Relationship time frame:** Many business-to-business transactions become ongoing relationships where trust is critical to maintaining a relationship over time.
- **Shorter Decision-Making Process:** You make decisions faster for a trustworthy company, shortening the purchase time span.

- **Brand Loyalty & Advocacy:** Happy customers are more likely to renew contracts, purchase additional services and recommend a supplier to others.

Trust Pyramid in B2B Sales

Trust in B2B sales is not created in a flash – it's earned little by little, layer by layer, just like a pyramid. This is the 'Trust Pyramid', with the layers building up from the bottom. Companies that develop credibility at every level of the pyramid will enjoy the genuine benefits of winning new business and maintaining long-term loyalty. Buyers need relevance, alignment, empathy and proof at every stage of a high-risk B2B buying decision.

Foundational Trust

The foundation of every B2B relationship is Foundational Trust, the bottom of the pyramid. This is simply to meet the minimum buyers' expectations before they even entertain the thought of you being a serious contender as a new supplier.

If what you're offering isn't a match with their business problem, or you come across as odd, trust will never have a chance to blossom. Building basic trust is demonstrating that you know their struggles and that you keep your word. Buyers must have a belief that your company does exactly what it promises – no surprises, no shifting. By being honest and dependable from the start, you lower the perceived risk of doing business with you.

- **Relevancy to Buyer Challenges:** Show that you get the client's industry and the issues they face. From that first interaction, you should start to customise your messaging and find a solution that comes down to what matters for them. A buyer will be much more inclined to trust a vendor who "gets" their problem and speaks their language.
- **Transparency:** Do not hide your product's features, pricing, or limitations from your users – they will see through that! Transparent communication instils trust. For instance, if there is any implementation difficulty or a learning curve, they should specifically mention it. This openness helps buyers feel you have nothing to hide.
- **Regulatory Compliance:** Ensure that you comply with any relevant

legal, security, and industry requirements (such as GDPR data requirements and any certifications in your industry). Privacy Compliance is the unspoken winner of trust – it gives you peace of mind that you won't be an unnecessary risk to do business with.

- **Consistency of response:** Deliver what you say you will. Send over the proposals, demos, services, etc., when you promise. Consistency is the sign of reliability. Over time, the small acts of delivery on small promises (such as actually calling someone at a scheduled meeting time or actually sending information you promised) add up to a larger impression: that you can be counted on when it counts.

Authority-Based Trust

Once the foundation is laid, we get to the second layer of the pyramid, which is Authority-Based Trust. So even if you've been clear, effective and trustworthy all along, savvy B2B buyers will still demand evidence in the form of third-party validation that it is the right decision to pick you. At this point, it's not just you saying, "Trust us"; it's other people vouching for you. You can gain authority through your industry reputation, expert endorsements, or examples of successes you've shown with other clients. The point is that buyers are just looking for proof that your company is a knowledgeable and competent leader in its industry, with a proven track record to prove it. And this corroboration from the outside also adds an extra layer of comfort. By displaying awards, certifications, and third-party testimonials, you increase confidence in your brand. You are no longer simply a vendor advocating on your own behalf; you're an established leader with the market's endorsement.

- **Accreditations / Credentials:** Point out any certifications, awards or standards that your business follows. Whether it's an ISO quality certification, a SOC 2 security attestation, or some sort of industry gold star, these badges are trust seals. They show that neutral authorities have sanctioned and checked you.
- **Third-Party Endorsements:** Use social proof from well-respected outside parties. For instance, analyst mentions (a reference from **Gartner** or **Forrester**), good press, or high ratings on B2B review sites can do wonders for your credibility. Hearing about your strengths from an

independent source will add credibility with suspicious buyers.

- **Client Testimonials and Case Studies:** Have your clients speak about your success. Publish case studies that demonstrate real-world outcomes you've achieved for clients, preferably those the potential buyer respects or can relate to. Tangible metrics (such as efficiency improvements or revenue growth) in these case studies show that you are reliable when delivering value. Similarly, testimonials or references from satisfied clients make you a business with a reputation.

Relationship-Based Trust

The top of the Trust Pyramid is Relationship-Based Trust, which results in genuine loyalty and perhaps even advocacy. That top level is reached only after credibility and authority proof is laid. The question is now not "Can I trust this vendor's promise and credentials?" but 'Do I trust that this partner can deliver our ongoing success?'. Here, relationship is key, and substance is the focus. Establishing relationship-based trust involves showing empathy, dedication, and interest in what is best for the client in the long term.

When a supplier approaches a customer's business as if it were their own, understanding needs, solving problems at the source, and always looking for opportunities to help drive efficiency, a high level of trust is developed. This is no longer about a product or about a contract; it's about a partnership. By showing empathy through actually listening and the ability to react to the customer's response, the customer is made to feel heard and cared for.

This never-ending focus on the customer ultimately matures trust into confidence and loyalty. The customer starts to use your business as more than a product or service; they use you as an adviser, a partner, and a part of their success. When trust in the relationship is high, customers stay. They renew contracts, grow their business with you aboard, and very frequently, you let them beat the drum for your brand. At the top of the pyramid is when a trusted vendor transforms into a trusted advisor.

A proactive engagement tailored to each customer: Dedicated Account Managers or Customer Success Contacts who really know the client's business. They should be customer champions, proactively checking in on specific goals, tailoring the advice given, and arguing the customer's case inside your walls. This one-on-one relationship adds a human layer to the business layer.

- **Exceptional Support:** Provide responsive post-sale support that makes customers feel safe and secure. This extends to rapid issue resolution, training and onboarding support, and check-ins to keep you up and running. The trust grows when a client trusts that you will help at any hour of the day or night. They get that you are more interested in their success than in closing a sale.
- **Ongoing Value Add:** Add value on an ongoing basis beyond just the initial sale. For example, share content about thought leadership, insights, studies, or best practices that can help the client succeed within their job or industry. You could offer exclusive webinars for customers or invite them to roundtable events. These are actions that let them know you care about them getting stronger.
- **Loyalty Benefits:** Acknowledge and compensate your customers over a period of time. Create loyalty programs or VIP opportunities that offer clients access to special benefits – access to a customer community, first dibs on beta programs for new features and executive briefings, are examples. These kinds of benefits feel rewarding, and they help promote a sense that being your customer is a partnership, not just a transaction. Customers who feel valued and a member of a special club trust in your brand more easily, and they are even more willing to advocate on your behalf.

From Trust to Advocacy

Once complete, the Trust Pyramid is more than just a satisfied customer, they become a loyal advocate. Foundational trust gets you in the door, authority-based trust seals the deal, and relationship-based trust keeps the partnership alive and kicking.

Thanks to you systematically covering each level with appropriateness, congruence, empathetic resonance and evidence, you are lowering the buyer's perception of risk and raising their trust. In the end, this deliberate building of trust is extremely hard to beat as a competitive advantage -

over time, anyway. Businesses that get to the top of the pyramid with their customers discover that those customers not only stay but also advocate on their behalf. In the B2B world, nothing is more valuable than a customer who trusts in you so vehemently that they become an advocate to your brand for others to use. This is the end game for building trust at the levels: multipliers of loyalty, word of mouth, and growth.

Managing Brand Reputation in the Digital Age

With businesses doing so much research online before they ever talk to a sales rep, brand perception is often driven by digital brand presence. You also need to monitor how you are being talked about on different channels.

Key Areas of Focus:

- Search Engine Visibility: Jobs with a high level of positive customer reviews, media attention, and analyst recognition are displayed as search results.
- Social Media Activity: Engaging/commenting on **LinkedIn** group discussions, answering questions, and sharing useful content.
- Online Reviews & Ratings: Asking your happy customers to drop testimonials on sites such as G2, Trustpilot and Capterra.
- Crisis Management & PR: Dealing promptly with unfavourable comments and keeping your brand story strong.

Trust as Purpose for Customer Retention

Trust builds not only when a deal closes; you must continually make deposits into trust to see returns in long-term partnerships and advocacy. Types of companies with success in post-sale engagement: You get better renewal rates and bigger upsells, and you get your growth through word-of-mouth referrals.

Post-Sale Trust Strategies:

- Lifelong Customer Training & Support: Continuing to provide resources that help customers make the most of their investment.

- Routine Performance Reports: Monitor KPIs and fine-tune when necessary.
- Community Building: Development of user groups, networking events and VIP forums for clients to interact and share their thoughts.

Trust: a Competitive Differentiator

Trust is not something you win once; it's something you work for, an ongoing investment that distinguishes business. Organisations that let transparency, credibility, and customer advocacy lead product promotion will win over those who depend on features alone. In the new B2B world, trust is the differentiation that shortens sales cycles and drives retention and long-term business success.

Post-Purchase Decision Reinforcement and Customer Retention

The B2B buying process isn't over when the contract is signed. The most important moment comes after the sale. Businesses not continuing to build confidence in their customers post-sale will see churn rates and poor word of mouth usually followed in the wake of 'un-up-sold/cross-sold' opportunities.

Why It's So Important to Reinforce After The Sale

There are several important purposes of post-purchase decision reinforcement:

- Avoids Buyer's Remorse: Good old' buyer's remorse is common for investments in B2B products/services. Reinforcing the decision reassures buyers that they made the right choice.
- Drives Product Adoption: The best products, even in the world, need a guidepost. Seamless execution leads to adoption and satisfaction.
- Builds Long-Term Loyalty: Urge customers to renew and buy again by bolstering your after-sale relationship.
- Increases Brand Advocacy: Happy customers are more likely to promote the service, write favourable comments and be used as case study references.

Understanding B2B Buyer's Remorse

Buyer's remorse is not only a consumer sales concern, it's a big problem in the B2B sales world, exacerbated by significant financial investment with the company-at-stake.

Typical causes of post-purchase remorse are:

- High upfront costs lead to over-anxiety over ROI.
- Frustration for the internal team due to convoluted implementation processes.
- Resistance by influencers in the buyer's organisation delays adoption.
- Sub-par performance if there's a disconnect between marketing promises and how the product performs in the real world.

Tactics to Build User Confidence Everywhere Before and After Purchase

To manage buyer's remorse and ensure customer retention, companies should take action to educate or reassure customers about their purchase decisions. Here are some of the tactics that help to drive post-purchase satisfaction and loyalty.

- Structured Onboarding & Training
- Offer new customers assigned onboarding specialists to help them through the implementation process.
- Provide role-based customised training to various users in your organisation.
- Create a knowledge base, how-to guides, and video guides to support self-learning.
- Hold interactive webinars and Q&A sessions to address concerns as they arise.

Proactive Customer Visits and Engagement

- Assign account managers to submit requests on the status and progress of customers periodically.
- Conduct quarterly business reviews to measure progress and find discipline for optimisation opportunities.
- Create automatic satisfaction surveys to gather feedback and catch problem areas early.
- Enable customer feedback loops and leverage the insights for product

offerings and service enhancement.

- Direct To Customer Communities and Networking
- Build private customer forums where users can share best practices and learnings.
- Host VIP events, industry roundtables and virtual networking sessions to further engage.
- Establish customer advisory councils to bring key customers in on future product plans.
- Publish comprehensive testimonies of customers' success, including what top users say as brand ambassadors.

Ongoing Education & Additional Value Resources

- Continue a consistent webinar series discussing advanced product features and industry trends.
- Provide certification training programs to augment customer skills and product expertise.
- Create benchmarking reports which enable customers to track performance gains over time.
- Write thought-leadership white papers around the latest issues and solutions in the space.

Community marketing (where you give yourself and your business a means to interact with your customers) is another way to help reduce the CAC (through referrals) and build trust as potential buyers see "normal" people promoting the product.
https://orangeowl.marketing/b2b-marketing/community-led-growth-in-b2b-marketing/

Using Customer Retention to Boost Growth

B2B success isn't about closing deals, it's about engaging, satisfying and keeping customers invested. New customers are 20 times more likely to stay when they trust you, and the cost of gaining a new customer is seven times higher than retaining an existing customer, so customer retention is a great growth strategy.

Retention Success Metrics:

- Customer Lifetime Value (CLV): Calculates how much money each customer is expected to bring in over time.

- Net Promoter Score (NPS): Measures how likely customers are to refer your company.
- Renewal Rates & Contract Expansions: Monitor retention and upsell opportunities.
- Customer Support Response Times & Satisfaction Scores: Indicates the level of service and general interaction.

Best Practices in Long-Term Retention:

- Issue customer loyalty programs featuring rewards for renewals and referrals.
- Personalise experiences by making recommendations based on past purchases and usage.
- Engage partners, drive early renewal discussions, showcase top-up services, and add value.
- Predictive support: The ability to address and mitigate potential pain points before they hit you.

Turning customers into Advocates

A good post-purchase journey is not just about keeping your customers; it's about converting them into your fans! Companies can foster long-term customer relationships that fuel future growth by investing in structured onboarding, proactive engagement, exclusive networking, and ongoing education. In today's competitive B2B space, post-purchase support is more than an afterthought; it's a major differentiator and a critical driver of long-term success.

The sale is only the start of the relationship with the customer. In B2B markets, where customer lifetime value exceeds its immediate value many times over, cementing the purchase decision and continued satisfaction is key to long-term retention and brand advocacy. Businesses that focus on post-purchase engagement will see higher renewal rates, more upsell opportunities, and greater customer loyalty.

Real-World Examples: B2B Buyer Psychology at Work

It's not just theoretical either, some of the world's most successful businesses employ B2B buyer psychology to generate revenue, establish trust, and shorten sales cycles. We have three recent case studies to show how brands have used cognitive biases, emotional engagement, and building trust to win over the decision between boxes and boxes of delicious cakes.

IBM's Thought Leadership Strategy: Tapping into Authority Bias
IBM has been a stalwart in the world of enterprise technology for decades, but its cloud computing, artificial intelligence and data analytics businesses are facing intense competition. To stand out, **IBM** invested heavily in thought leadership and educational content centred on acting as an advisor rather than a visiting vendor.

Psychological Tactics Used:

- Authority Bias: **IBM** establishes itself as an expert by providing comprehensive white papers, research reports, and expert analyses in leading-edge areas like AI, cyber security, and quantum computing.
- Social Proof – **IBM** often cites work with leading global organisations and governments, strengthening their social proof.
- Risk Aversion you can Hedge against: **IBM** alleviates the buying company risk by providing ample evidence that an investment in the technology is safe.

Findings: Through reports and industry insights, **IBM** is a go-to source for C-suite executives, with 69% of buyers agreeing that high-quality thought leadership increases their trust in a brand. https://insightsmedia.co.uk/articles/thought-leadership-content-good-for-brand-perception-and-sales/

Salesforce Using Social Proof and Community Engagement
By being customer-centric, faithful, and emotional, **Salesforce** is a leader in the CRM software industry and helps to grow a huge ecosystem. The business knows that B2B buyers need peer validation before they pull the trigger.

Psychological Tactics Used:

- Social Proof: **Salesforce** has thousands of customer successes written up about businesses across various industries that have benefited from their solutions.
- The Bandwagon Effect: With its yearly Dreamforce conference, **Salesforce** assembles tens of thousands of business professionals in an effort to position **Salesforce** as the de facto CRM service.
- Community-Driven Trust: **Salesforce** developed Trailblazers, an online learning community where customers exchange ideas, support one another, and feel confident in their decision to use **Salesforce**.

Results: Dreamforce is now one of the largest and most influential B2B events in the world, drawing over 170,000 attendees each year (pre-covid numbers) and generating unmatched brand advocacy.
https://www.businessinsider.com/5-things-to-know-before-salesforces-tech-conference-dreamforce-2017-10

Businesses that interact with Trailblazers have greater product satisfaction and retention because customers can feel they are a part of the brand and the community.

HubSpot's Freemium Model: Reducing Perceived Risk

We've seen what **HubSpot**, a pioneer in inbound marketing software, has done by using Freemium to combat loss aversion and risk perception in B2B software purchasing.

Psychological Tactics Used:

- Loss Aversion: By giving away a free version of its CRM, **HubSpot** removes the monetary investment, so companies have nothing to lose by trying the product out.
- Anchoring Bias: **HubSpot** places its free CRM as the foundation, which presents its paid tiers as high-value add-ons.
- Commitment & Consistency Principle: Companies who start using the free CRM and build it into their processes are more likely to commit to paid editions for more features.

Outcomes: **HubSpot**'s freemium model has resulted in higher customer acquisition rates, with users converting to paid plans at a much higher rate than traditional sales-led approaches.
https://www.chiefcustomerofficer.io/newsletters/the-bet-that-hubspot-made-in-2014

Lessons Learned from these examples
Thought Leadership Builds Trust: Brands such as **IBM** leverage authority bias with some research-based content to become the see-as leaders of their industry.

Social Proof & Community Building Creates Brand Loyalty: Witnessing Salesforce's streak in success, no one can deny the role of Customer Advocacy, Events, and Peer Validation.

Lowering the Risk: Low Barriers to Entry - **HubSpot**'s freemium model demonstrates how reducing perceived financial risk speeds up adoption and boosts revenue.

B2B marketers who leverage these principles in our current landscape will not only win new customers, but they'll also convince them to stay, engage, and advocate, a winning equation to win in today's noisy world.

Mastering the Psychology of B2B Buying

B2B marketing is not just about showing the product features or pricing competitively; it's about understanding the way the buyers think, feel, and decide. The psychological implications of B2B purchase decisions, from cognitive biases to emotional triggers to trust-building strategies, have a radical influence on whether you book the sale or lose out.

Firms that build psychology, risk reduction, trust, and emotions into their strategies will have increased conversions, a reduced sales cycle, and a more loyal customer base.

3. Charting the Modern B2B Journey

"The best way to predict the future is to create it."

Peter F Drucker, Grandfather of Modern Marketing.

Understanding the B2B Buyers' Journey

The B2B buying journey is complicated, convoluted, and hard to predict. Even the slightest B2B sale is long, requires numerous players, an investment of a decent sum of money and includes a long-term contract. In high-stakes investments, decisions may take months or even years while companies weigh options, evaluate risks and bring in multiple departments and decision-makers.

The business landscape is complex, involving research, strategic thinking, and consensus building. B2B purchasers aren't just after a product or service; they are looking for a long-term relationship that will add value over time, help solve complex problems and align with their own strategic objectives. Therefore, B2B marketers must accept that they are selling to a complex decision-making arena that extends beyond simple sales. To win in this environment, marketers need to think of the buyer's journey as a marathon, where brand presence, credibility, and trust matter as much as product functionality and competitive pricing. There's a lot to it; here's an insight from **McKinsey:**
https://www.mckinsey.com/capabilities/growth-marketing-and-sales/our-insights/improving-the-business-to-business-customer-experience

The "Iceberg" Analogy of the Buying Journey

An excellent analogy for the B2B buying journey is that of an iceberg. Like an iceberg, most of the decision-making process takes place "below the surface," out of sight of potential suppliers. Prospects spend much of their buying journey doing their own research, having internal discussions, and exhaustively researching their options long before they even think about contacting a vendor. According to Gartner, B2B buyers complete as much as three-quarters or more of their journey before ever reaching out to suppliers. https://www.gartner.com/en/sales/insights/b2b-buying-journey

Buyers frequently need to articulate their needs clearly, vet their options, and build internal support before they ever speak to a vendor. They only start contacting their suppliers when they get towards the end of their withdrawal process, and at this point, they usually have a very set notion of what exactly they are looking for.

Invisible Research and Evaluation

The hidden aspect of the B2B journey poses a unique challenge for marketers, who must ensure that their brand is consistently visible and credible even when potential buyers aren't directly engaging with them. During this stage, companies need to be able to attract the attention of buyers: this requires investing in a strong digital presence, developing content that positions you as an authority, and staying visible in industry channels.

In this invisible stage, potential buyers are comparing vendors and solutions online without speaking with anyone. It could be that they are visiting your website, reading reviews or comparing case studies, engaging in your thought leadership content or leveraging other types of content. So, it's vital to have resources in the pipeline, such as educational blog posts, white papers, videos, and case studies that provide answers to potential questions, shake out objections, and continually prove your credibility. By predicting the data that buyers require and ensuring they can find it, marketers can influence the 'hidden' part of the buyer's journey, making sure that when prospect do raise their hand, they've already heard positive things about your brand and your work.

Marketing's Challenge

The problem here is that the B2B marketing team need to make an impression and get ahead of the competition before these buyers venture out of this 'cloaked' research period. The role of marketing is one of credibility and visibility: establishing a brand as a reliable and expert resource that buyers will seek out when they are ready to contact. This calls for a hands-on strategy for producing content, promoting thought leadership and branding building to ensure your brand is respected, top-of-mind and easily identifiable when potential buyers start their research.

Building this presence means being active on various outlets – social media in all its permutations, blogs, industry publications, and trade shows. Through the construction of a complete ecosystem of content and brand assets, marketers can "own" a category they exist in, and buyers have an easier decision to make when the time has come to act in selecting those products or solutions.

The Non-linear B2B/Purchase Process

Many Buyer Journey charts show a clear progression from awareness (the top) to, let's say, purchase (the bottom), but actual Buyer Journeys are often not that neat. The B2B purchase journey is seldom ever a simple one. Instead, it is often a non-linear process, with both buyers and previous decision stages moving back and forth, stopping, returning to earlier states, and even changing direction altogether as priorities change. The multiple changes, insights, or new issues that can interfere with the scheduled buying process can make the route taken look like a haphazard journey.

Adapting to a Non-linear Path

The non-linear nature of navigation means that you need to span a longer time horizon as a marketer and aim for relevance, relationships, and flexibility rather than the quick and dirty converts. B2B marketers have to realise that leads can still come back to the brand after months (and sometimes even years), specifically when it comes to big decisions and/or high-value purchases.

The intricate nuances of the B2B journey demand that marketers practice flexibility and patience during this process. Rather than herding buyers to a faster conclusion, good marketers keep the brand top-of-mind, helping buyers arrive at a decision point with a better feeling about the brand. This long process engagement is key to B2B, where buyers' requirements and preferences change over time and where the journey often involves long pauses and sudden changes, of course.

Breaking Down the B2B Buyer's Journey Model

The B2B buyer's journey can be highly variable from organisation to organisation, industry to industry, and case to case, but most models distil the process into six major stages that can serve as a helpful guide to the journey – from first-time awareness to how one engages post-purchase. By matching marketing initiatives to suit each stage, B2B organisations can go to where their buyers are and present relevant content that can help ease their way through the process.

Stage 1 Awareness of New Needs

The process starts as prospective buyers become aware of a problem or a need. In the early stage, this is typically triggered by internal difficulties, changing markets, regulatory transformation or growth ambitions. At this point, the buyer might not exactly know the solution they need, but they know something needs to change.

Marketing's Role

During the awareness stage, marketing aims to create visibility and to become recognised as a thought leader. Rather than sending a particular product solution, you need to create content that speaks directly to the buyer's challenges, shows you understand their industry, and positions your brand as a credible source of insight. You could publish blog posts, report industry news, schedule webinars, and release educational content that demonstrates common issues and new trends.

Through sharing valuable information and not pushing heavy sales, marketers can begin to capture the attention of buyers who are at the very start of their journey, raising awareness without forcing engagement. This generates trust and credibility, as buyers are now more likely to look at your brand when it is time to progress to the next stage.

Stage 2: Options Assessment

How it works: After a pain becomes real, buyers go to the marketplace to research solutions. This discovery phase consists of profiling competing products or services, studying their features and capabilities, and short-listing them. Buyers want to know what's out there, and they need to understand which solutions are best suited for solving their unique challenges.

Providing Educational Content

Simply put, advertising is highly targeted, along with educational content that showcases a newer brand's unique features while making it stand out from the competitors. Here's the time to feature case studies and client success stories, address frequent Q&A and offer deep guides that assist buyers in their purchasing journeys. When you position the brand with something of value, something you know they have in common, you can start building trust and generating interest in solutions.

Informative, educational content is particularly useful at this stage. Purchasers value tools that allow them to compare options and define decision-making criteria. By providing deeper context on what your product can do, how it's rolled out, and what kind of potential upside or improvement they can expect, your marketing, in essence, gives prospects permission to take you seriously.

Stage 3: Removing Risks and Concerns

As purchasers move down their buyer's journey and start to narrow their choices, they typically move toward risk reduction. In B2B, purchases represent a considerable investment of time, resources, and, of course, money, so stakeholders are understandably wary. Their primary focus is they want to know the solution they select is not going to let them down and is going to work for them in the long-term.

Establish Your Credibility and Lower Their Perceived Risk

At this point marketing's job is to establish credibility and assure the buyer that they are making a good decision. You can do this by sharing client reviews and testimonials, publishing case studies that show proof of your work, drawing attention to endorsements from third parties and displaying your certifications and awards. The objective is to present the brand as a low-risk, safe purchase that will enable prospective buyers to move forward with confidence.

Risk mitigation is crucial in B2B because mistakes tend to have higher stakes. By leveraging transparency and showcasing proof of success, along with what makes them different from their rivals, marketers can reassure potential clients and make a strong argument for them to choose their product or service.

Stage 4: Decision

People who are in the decision stage are often close to making a vendor selection and might just be looking for some last input from a director or other senior buyer, procurement or another decision-maker. They now need a little push to convince them that they're going to make the right decision, and they might have a few questions or need some extra confirmation to go ahead.

Supporting the Final Decision

Marketing and sales need to collaborate very closely in this stage to give the buyers the information and support that they need to make their decisions. This could be anything from a product demo and ROI calculator to a full proposal and personalised consultation. By answering the last few objections and providing personalised direction, marketers and sales can take the path to confident, educated decision-making.

The decision stage is an important phase in which educators and sales can have a huge influence working together. And working in tandem with your team and each other, they can ensure a seamless, supportive path so that buyers arrive confident in their choices, minimising the chance of last minute jitters, holds or even 'Consensus Fatigue' where buyers disengage because the internal process is too complex!

Stage 5: Implementation of the Purchased Solution

When the decision has been reached, implementation of the chosen solution will be the primary concern. This phase is also where it is most critical in establishing a positive relationship with the client, as a well-coordinated implementation can promote confidence in the brand and serve as a good base for future customer satisfaction.

Post-Sale Engagement

Marketing is not done once you've sold your product; you need to stay connected to the customer through the implementation process. Frequent biweekly, weekly or monthly progress check-ins, tutorials, onboarding resources and training sessions can all help to make sure the product or will have the effect we told them it would and solve their problems. By providing continued support, marketers can enhance the client's satisfaction and loyalty, which may lead to repeated business and word-of-mouth referrals.

Stage 6: Developments Over Time

While customers continue to use the product or service, they may uncover new needs or identify improvement opportunities. This creates an opportunity for upselling, cross-selling, and building long-term relationships.
Advocacy has something to say about staying relevant by proactively offering information on new features, updates and other services that go with the product. By showing longevity and a true desire to be part of the client's success going forward, the brand becomes a partner who is determined to help the client grow and change.

Marketing-Led vs. Sales-Led Stages of the Buyer's Journey

At various stages along the B2B buyer's journey, different strategies are required of your marketing and sales departments. Broadly, the early stages, awareness, assessment, and risk reduction, are marketing-heavy, and the later stages, decision, implementation, and adaptation, are sales-heavy. However, the most successful B2B strategies will see these roles working hand in glove to provide a joined-up and cohesive experience for the buyer. The more you know which phases are marketing led and which are sales-led, the more both teams can coordinate and reinforce, which in turn maximises your impact and keeps your company's message in perfect step with the buyer's shifting preferences.

Phases Led by Marketing (Stages 1–4)

The first four phases of the buyer's journey, awareness, assessment, risk reduction, and decision, are dominated by marketing activities. At the early stages of this time window, prospects are still researching, considering their options, and weighing possible solutions. For marketers, this is a chance to elevate the brand as an insightful authority by making demands, cultivating leads, and building trust.

Demand Generation and Thought Leadership

In the beginning, the focus of marketing is to build demand and establish the brand as a leader in the market. This means creating copy that catches eyes, educates consumers, and gets them more invested in whatever you have to offer. Standout content types in this stage are white papers, blog posts, webinars, infographics and industry reports. By offering real insights, marketers can capture the attention of prospects and illustrate the brand's expertise.

Thought leadership is of special importance in B2B marketing when buyers are looking for solutions that are informed and contain industry knowledge. Thought leadership: Longer thought leadership-style content pieces, like in-depth research reports, interviews with industry experts and insightful articles, are a way to build credibility and gain the trust of your audience as a marketer. It's not just a product-out content but a window into what the company knows and believes in and how it is in service of its customers' success.

Lead Nurturing and Relationship Building

Once the early demand is created, it is all about developing those leads and relationships. The lead nurturing process is slow, and it involves feeding prospects with that which seems necessary at the end part of the journey hanging necessary for them. This could be things like targeted email campaigns, follow-up assets, customer case studies, and webinars. When marketers deliver regular, relevant content, it keeps the brand front and centre and nudges prospects along the path to purchase.

Lead nurturing is also dependent on developing a trusting relationship. In place of making the hard sell now, marketers must make an effort to learn what each prospect needs and even wants, then use that intelligence to inform you with the best help and support. Good lead nurturing develops a sense of trust and increases the probability the lead will choose to use the brand in the future.

Sales-Led Phases (Stages 5–6)

When buyers progress to the decision and implementation phases of the buying journey, the sales organisation takes over, helping prospective buyers make final assessments, negotiations, and the buying decision. Now, while the emphasis is not on education and awareness to hands-on support, personalised support to ensure that buyers feel comfortable in their decision.

Tailored Support from Sales

As potential purchasers get closer to making a decision, they tend to need more specific, personalised information that they can use to answer specific questions and resolve specific issues. And here is where the sales department comes in, with personalised consulting, product demos, ROI discussions and answers to objections in a direct conversation. By collaborating with the prospect, the sales team are able to create a stronger, more personal link with the buyer, thus making the buyer feel more important and cared for after the purchase.

Sales professionals must know the background and insights that marketing has shared so far so that they can work from that to keep the narrative going. That seamless handoff from marketing to sales makes for a smooth and consistent buyer experience, enhancing the perception of trust and reliability of your brand.

Ensuring a Seamless Handover from Marketing to Sales

A smooth handoff from marketing to sales is one of the most important hurdles in B2B marketing. Disconnect between these two groups can create a poor customer experience and drive a wedge down the sales funnel. To avoid them, marketing and sales need to come together, sharing information and communicating openly.

Here's how marketing can lend a helping hand to the sales team by developing assets that speak to the common objections, available ROI and the product's differentiator. These could be custom case studies, features to compare competitors and detailed product specs. When sales agents are armed with the things they need to answer the buyer, they are better able to facilitate the buyer's journey from interest to buying.

The Role of Lead Nurturing And Long-Term Engagement

Few leads convert quickly when you do B2B marketing. B2B buyers typically require sustained engagement over long periods of time before they're prepared to purchase. This fact makes lead nurturing a critical part of any successful B2B plan. Lead nurturing is therefore based on fostering a relationship of trust, relevance and value whereby prospective leads consider the brand a reliable partner, even if they are not immediately ready to purchase.

Persistence and patience consistently are worth their weight in gold within high value B2B SaaS sales – in this type of complex sale, decision making is spread across and between many stakeholders with deep and detailed due diligence taking place. These leads need to be nurtured, and this will require a constant effort to deliver value, understand why buyers buy what they buy and show up for them every step of the way.

Lead Nurturing Strategies

Lead nurturing that actually works is more than just getting a series of emails in front of a contact; it's a meaningful experience that changes based on the interests/needs/stage in the buyer's journey of each individual prospect. Lead nurturing might take the form of email sequences suited to where someone is in their funnel, retargeting campaigns with a personalised element, or the provision of segmented content that addresses someone's particular problems or queries.

Personalised Engagement

The key to successful lead nurturing is personalisation. In the B2B arena, where decisions are big and complex and apply to a very particular scenario, the level of generality is way off the mark. By personalising every touch for the individual's needs and interests, marketers are able to form a richer, more meaningful relationship. That might mean slicing audiences by criteria such as industry, job role, pain points, or past interactions with the business.

For example, if a lead has expressed interest in a product feature or attended a webinar on a topic, marketers can follow up with case studies about that feature or more resources that cover the topic. This personalisation shows that the brand is listening, understanding, and empathising with the prospect's challenges, which deepens the level of trust and increases the likelihood that they will make it a habit to turn to the brand when they are ready to make a decision.

ABM: Engaging the right accounts with account-based marketing

At its most strategic level, Account-Based Marketing (ABM) focuses on the personalisation that inbound encourages even further by considering individual accounts as unique markets. Instead of a broad-based approach, ABM is a targeted marketing approach where, rather than casting a wide net, you are zeroing in on a specific niche of high-value accounts with very personalised marketing activities. It's especially valuable in a B2B, complex sales environment where targeting individual companies and the decision-makers within them can make a huge difference.

Strategic Plans on Key Accounts

Within an ABM approach, marketing & sales collaborate to produce tailored campaigns for their target high value accounts. This could include custom content, private webinars, personal demonstrations of the product, and conversations with the decision makers. By coordinating closely with sales, ABM campaigns are intended to reach and speak to specific people within a specific company and to answer their needs, questions, and concerns accurately.

If you're focusing on a much larger financial services company, your approach around account-based marketing might be to develop a series of white papers that speak to the issues that are unique to the finance sector and help people understand that your solution helps you deal with industry regulations or data privacy. With content that speaks directly to the needs of a specific account, ABM generates a high degree of relevance, enhances engagement, and helps a brand to cut through the noise of a crowded market.

ABM programs tend to require a robust knowledge of each account's pain points, aspirations, and industry context. It's the kind of personalisation that takes time and resources but also the kind that can make an impact on average monthly personalised touchpoints, average number of days to close, and improved relationships with key target accounts. Refer to Chapter 5 for more information about ABM & ABX (Account Based Experience) the future form of ABM.

B2B marketers also have a huge responsibility to instil trust in your target audience by providing relevant messaging consistently and showing your brand as an authority figure these should be you and a resource that your audience will trust. Trust is a pre-requisite to customer relationships, and high value purchasing decisions need "deciders" to be assured of these honest, open and reliable relationships with their vendors.

Brand Consistency and Messaging

You don't break trust by being inconsistent in your messaging; we can probably all agree on that. Prospects should see you everywhere: in blog posts, in webinars, in case studies, and in industry publications. When they do, they should hear the same message, the same tone, and the same level of professionalism. It is this consistency that upholds the buyer's confidence that the brand is solid, reliable and will live up to its promises.

Consistency is not just about what you say, it's how your brand acts, looks, and feels across every touch point where your buyer interacts with it.

Consistent brand identity across a website, social media accounts, and email communication enhances the credibility and professionalism of a business. In branding terms, the consistency of the offering helps consumers identify the brand and remember it, which makes the buyer feel as if he or she can trust the brand or be comfortable with the brand enough to make the purchase.

Consistent messaging is not just about how channels look the same. It also takes a brand staying true to its values and mission. By communicating a consistent brand message, companies cement their credibility and reliability, becoming trustworthy partners that buyers can depend on in the long term.

Thought leadership content could be long blog posts, industry reports, white papers, podcasts, and expert interviews. The idea is we need to give so much more so that it's providing value, answering questions and innovating on industry trends. When done properly, thought leadership builds trust and credibility, as prospects come to trust the brand as a trusted source for knowledge and quality insights.

Measuring Success Throughout the Buyer Journey

Monitoring the performance of a B2B marketing plan is vital to making it better over time. Marketing teams can measure their impact, see where they are strong, and learn where they are weak by tracking KPIs (key performance indicators) at all stages of the buyer's journey. The metrics of success need to extend beyond conversions and to measure the extent and depth of engagement, the quality, and the journey the leads take along the way.

Key Metrics to Track

There are specific metrics for each stage of the buyer's journey that can help you understand how well your marketing is aiding prospects in moving along on their journey towards purchase. Here are some important measurements to track at each level:

Awareness Stage:

Basic KPIs are brand mentions, website traffic, social media engagement, and content shares, which are all clear-cut signs of the beginning of the interest mode. They demonstrate how well the brand is getting the attention of its target market and creating the necessary awareness.

Evaluation:

This is where a buyer is researching and consuming educational content. Content downloads, webinar attendees, and time spent on product pages are key metrics to track. These measures are the quantum of the reach of the brand in assisting buyers to understand their needs and explore possible solutions.

Risk Reduction Stage:

At the risk reduction stage, potential buyers are trying to decrease perceived risk. Metrics such as case study views, testimonial engagement, clicks, and engagement with third-party endorsements give an indication of how effectively the brand is addressing doubts and gaining trust.

Decision Stage:

In the decision stage, buyers are ready to slay their decision-making dragon. Indicators such as demo requests, ROI calculator usage, and the number of sales-qualified leads (SQLs) progressing into the proposal phase capture momentum toward a purchase and the impact of decision-support materials.

Implementation:

Customer attitude scores, on-time implementation scores, and feedback from onboarding calls/meetings help to evaluate the success of the implementation and also pinpoint some areas that need improvement.

Adaptation Stage:

As part of the long-term engagement stage, metrics like customer retention, repeat business and upsell/cross-sell conversion rates provide insight into the effectiveness of repeat support and how well your brand adapts to changing customers' needs.

Tracking such data would enable B2B marketers to see a holistic view and optimise their digital and marketing strategies accordingly. Frequently updating these stats and metrics allows for data-driven pivots so that a team's approach is always fresh, effective, and relevant to their organisation's needs.

Analysing and Adapting for Continuous Improvement

While metrics are especially valuable for giving us insights overall, metrics are important in how they're used for improvement. Keep a close eye on performance data and monitor it to identify patterns and trends, as well as areas which may need tweaking. For example, if engagement is high when the prospect has increased awareness but falls off during consideration, it may be a signal that the need is for in-depth, educational content that will allow the prospect to evaluate the options.

We all hear that practice makes perfect, and continuous improvement is certainly a fundamental component of fruitful B2B marketing. By harnessing data to pinpoint where to plug holes and where to better replicate content and lead nurturing strategies, marketers can develop a more effective and agile strategy that skews to the shifting landscape of the market. This agile methodology ensures that B2B marketing is always fresh, effective and in tune with what buyers want.

It is very important for a successful marketing strategy to know the B2B buyer journey and create a model around it. The process is usually not a straight-line process but usually a non-linear, continuous internal discussion and interaction on different touchpoints. By tailoring marketing to each stage: awareness, assessment, conclusion, implementation and adoption, businesses are able to build a model that they can use for commerciality to grab the buyers' interest and ensure they do not lose momentum, which results in a better conversion rate.

With the current competitive nature of B2B relationships, lead nurturing and relationship building for the longer term are more critical than ever. What lead nurturing is really doing is building trust, providing value, and positioning the brand as a reliable partner.

Trust is fundamental in B2B relations. By being consistent and credible in your messaging, investing in thought leadership, and sticking to your natural market, B2B marketers can build a platform of trust that will get potential clients to select your company when they're ready to pull the trigger. Thought leadership, in particular, is a great way to gain credibility and help the brand position itself as the leading authority.

Finally, if success is measured by data at each step along this journey, this leads to continuous improvements. Through the ongoing analysis of KPIs and data-based tweaking, B2B marketers are able to optimise their approach, content, and customer experience. By embracing an integrated model of the B2B buyer's journey, businesses can break free from a collection of disjointed tactics to create a unified, sustained agenda that will result in real growth. Through relationship development, trust, and data-informed understanding, B2B marketers can establish their brand as something more than just a vendor - a partner in their customers' success.

Summary:
An adaptive marketing framework is needed for this increasingly complicated landscape. This model combines customer intelligence, content strategy, multi-channel engagement and measurement.

- Develop a marketing model that can keep pace with changing customer habits.
- Pay attention to why customers are doing things at each step in the framework.
- Always be testing and improving activity.

Case Study: IBM's Transformation of the B2B Buyer Journey
IBM, a global leader in technology and consulting, recognised the shifting dynamics of the B2B buyer journey. With buyers conducting extensive independent research before engaging with vendors, IBM aimed to realign its marketing and sales strategies to meet these evolving behaviours. IBM implemented a comprehensive strategy focusing on the following key areas: Content Marketing and Thought Leadership, Digital Engagement and SEO, Personalised Nurturing Campaigns, and Sales and Marketing Alignment.
https://medium.com/venturehq/ibms-content-strategy-an-instructive-case-study-in-b2b-marketing-b8818e91b282

4. Bridging Silos: Uniting Sales and Marketing for Growth

"It's not what you sell that matters as much as how you sell it."

Brian Halligan, Co-Founder of HubSpot.

There has been a fundamental change in how B2B businesses think about sales and marketing alignment. Gone are the days when sales and marketing worked in silos, with little more than a "gut feeling", past performance and manual mechanics to inform their engagement strategies. The new reality that AI is ushering in is automation, better data-driven decision making, predictive decision-making, and hyper-personalised customer engagement.

AI is way more than a better tool; it's the engine powering a new species of business agility, but only when combined with strategic integration. There is an acceleration for those companies that are using AI's leverage with their sales and marketing processes, and the risk of being left behind for those clinging to old ways. Today's buying path is more complicated than ever, with several decision-makers, digital touches, and extended decisions. Sales/Marketing needs to sing together in perfect harmony, so step into this level of orchestration carefully; in the past, that's gone very, very wrong. And in case you're wondering, let's imagine AI is the baton that makes sure the symphony can navigate all the complexity.

How AI is Changing the Way People Buy and Your Customer's Expectations

According to **LinkedIn**'s State of Sales Report 2023, 87% of sales and marketing leaders believe alignment between their teams enables business growth, while organisations with well-integrated teams see 209% higher marketing revenue contribution.

https://www.lxahub.com/stories/sales-and-marketing-alignment-stats-and-trends-2023

These incredible statistics assert the need for dismantling traditional silos and promoting AI-driven cooperation.

AI-powered tools help companies to not only be present when these self-learning buyers come up but also to engage them proactively, by providing personalised content, predicted recommendations and real-time insights based on their interests and needs.

And if your message sounds generic, your buyer has already tuned out.

Hyper-personalisation is no longer a 'nice to have'. AI-powered customer-relationship management (CRM) systems, automated chatbots and predictive analytics tools allow businesses to predict their customers' requirements, offer the right solutions at the right time and remove friction from the buying process. This is rapidly transforming the face of sales and marketing, and consequently, sales and marketing professionals are now operating in a world where prospects no longer tolerate token or generic messaging.

Artificial Intelligence Alignment is the Future of Sales and Marketing

Unlike old-school alignment, where sharing even stripped-down data is still done manually and coordinating even simple campaigns requires some heavy lifting, today's alignment is AI-powered, data-first and ever-evolving. It's no longer sufficient for sales and marketing to meet periodically to exchange insights, this alignment needs to be instantaneous, dynamic and based on AI-driven analytics. If you're still treating alignment as a quarterly catch-up, you are already behind.

Sales professionals require real-time customer intelligence so they can interact with prospects at the right time with the right message. Marketing departments then need to go one step further and fine-tune their targeting, as they use AI-driven audience segmentation and predictive lead scoring to ensure they're appealing to high-intent individuals. And when you augment them with AI, the two functions act as a unified ecosystem rather than in isolation.

Think about what AI-powered revenue attribution can do. Businesses employing AI-based models that can analyse the customer journey across a range of touchpoints can pinpoint which campaigns, content, and sales strategies most directly drive conversion. This takes the guesswork out of the equation, enabling businesses to assign resources properly and maximise engagement activities for the best possible ROI. Without AI, sales and marketing alignment is reactive; with AI, it is proactive and predictive.

This chapter will examine the fundamental drivers of sales and marketing alignment, the influence of AI on accelerating collaboration, some real-world companies that have adopted AI-powered alignment, and the first steps you can take if you're interested in a tighter business-wide collaboration. It will also highlight the impediments to effective alignment and how AI can remove those obstacles to generate more leads, attribute more revenue and create better customer experiences.

Why We Need AI-Driven Sales & Marketing Alignment

It is no secret that in the past, sales and marketing teams were isolated from one another and that misaligned goals would lead to mixed messages and an untidy handling of leads. These gaps act as a barrier to revenue growth and make your teams less effective, which means lost revenue opportunities, annoyed people, and squandered resources. It can be challenging in case of a mismatch: marketing creates leads that are not accepted by the sales department, and the sales team cannot close because there has been no proper marketing engagement in advance. Imagine a sales team working leads from a webinar while the marketing team celebrates white paper downloads, but no one joins the dots. If you don't have alignment, then you have friction disguised as progress.

Companies with aligned sales and marketing teams have 32% more company revenue, 36% more organisational customer retention, and 38% higher win rates, according to RenderTribe.

https://www.rendertribe.com/the-key-to-a-greater-roi-sales-and-marketing-alignment/

This dynamic is changing, however, as we move toward AI-based collaboration. Classic alignment was based on human interaction, physical meeting locations, and siloed tools, which often resulted in inefficiencies, delays, and general confusion. AI eliminates these bottlenecks by facilitating data-driven, real-time insights, even predictive analytics, and automated workflows, thus making sales and marketing departments an interconnected ecosystem rather than two separate divisions of the same company.

Buyer expectations: the new reality?

B2B buyer behaviour has changed significantly. Modern buyers value conducting their own online research; they are already 57% through the buying process before ever engaging with sales, so it's vital that marketing provides the right content at the right time to assist buyers along the journey. A whopping 70% of the B2B buyer's journey is already done by the time prospects are willing to speak with a salesperson.
https://6sense.com/blog/dont-call-us-well-call-you-what-research-says-about-when-b2b-buyers-reach-out-to-sellers/

This shift means that sales teams rely heavily on marketing-driven engagement, yet without proper alignment, these efforts often fail to resonate with high-intent buyers. AI allows for accurate buyer journey tracking so that marketing knows they are delivering content at exactly the time sales receive qualified leads who are truly ready to engage. Furthermore, AI enhances customer intelligence, allowing companies to predict pain points, objections, and the likelihood of conversion with unprecedented accuracy. Companies leveraging AI-powered sales and marketing strategies can outperform the competition by providing hyper-personalised, data-driven interactions across the entire customer base.

How AI is Transforming Sales and Marketing Alignment

Traditional alignment focused on manual data sharing and simple campaign coordination stands apart from AI-driven, data-rich, persistent new alignment. The sales team should have the ability to trigger an interaction at the ideal time and communicate with the correct message at the right time. Marketing teams need to sharpen their targeting further, leveraging AI-powered audience segmentation and predictive lead scoring to bring in high-intent prospects.

Imagine the power of AI-based revenue attribution. Companies that leverage AI to monitor customers throughout the various touchpoints will see which sets of campaigns, which types of content, and which sales tactics influence conversion the most. This removes all the guesswork, which makes it easier for businesses to best allocate resources and focus their engagement efforts for the highest possible return. Without AI, sales and marketing alignment is reactive; with AI, it becomes proactive and predictive.

By embedding AI-driven intelligence into their sales and marketing processes, companics can track the alignment of both teams as a synergistic, mutually supportive unit, not a siloed couple of different bodies. This chapter will cover the most important factors influencing sales and marketing alignment and how AI helps improve this situation while showing use cases of organisations that successfully adopted AI-driven alignment strategies. It will also focus on points of friction that are hindering successful alignment and how AI can work to remove them to maximise lead generation, revenue attribution, and the customer experience.

Key Benefits of AI-Driven Sales and Marketing Alignment

Enhanced Customer Experience

AI-driven personalisation means your prospects are introduced to content and recommended resources which are specifically relevant to their pain points, which in turn captures their interest and trust.

- Artificial intelligence chatbots offer live assistance and responses, suggesting relevant options and holding the hand of any customers through a predefined funnel.
- In sentiment analysis, we measure how customers feel, analysing their tone and behaviour and reflecting that in our messaging to make things more personal and human.
- Dynamic content personalisation uses AI to shape website experiences, email content and ads to give a sense of relevance and context to interactions.

Higher Lead Conversion Rates

Artificial Intelligence based predictive analytics also optimises lead scoring rules and allow sales teams to target only high-value prospects with a higher likelihood of converting.

- AI monitors behavioural triggers (e.g., website visits, content engagement, email open rates), identifying when a prospect is actively considering a purchase.
- Automated follow-ups powered by AI ensure no leads slip through the cracks, using optimised timing and message personalisation to nurture leads more effectively.
- Machine learning models refine lead segmentation, allowing businesses to prioritise prospects based on historical conversion patterns and engagement levels.

Shortened Sales Cycles

AI powered automation removes bottlenecks in the lead nurturing process, lowers friction in the sales funnel, and increases the velocity of deals.

- Automated lead qualification methods immediately recognise high-intent prospects so that sales team members speak only to the most viable leads.
- AI enhanced CRM tools suggest the best time for outreach based on a prospect's digital interactions, ensuring that sales reps engage at the optimal moment for conversion.
- Conversational AI handles initial outreach, answering FAQs, qualifying leads, and scheduling meetings, allowing human sales reps to focus on high-value interactions.

Higher Revenue And Marketing ROI

Organisations that use AI to align Sales and Marketing see a 29% increase in sales growth. Here's how AI makes marketing more efficient by improving targets, automating follow-ups, and eliminating wasted ad spend on low-intent leads. AI-based budget allocation means that marketing spends are efficiently optimised for impact, focusing on the most effective channels and campaigns.
https://www.prnewswire.com/news-releases/revenue-organizations-using-ai-in-2024-reported-29-percent-higher-sales-growth-than-their-peers-according-to-new-report-from-gong-302312896.html

- Sophisticated attribution models follow many touchpoints and determine which of your sales and marketing activities really helps the revenue so we can make smarter investments.
- Highly personalised outreach tactics lead to better engagement levels, greater lead quality, and cost efficiency for customer acquisition.

Real-Time, Data-Driven, Decision-Making

Among AI's most impactful powers is its capacity to both crunch huge troves of data and deliver real-time analysis so sales and marketing professionals can make faster, smarter decisions.

- Marketing campaigns with high ROI can be identified using machine learning algorithms so businesses can push their chips into what's working best.
- Predictive AI forecasts demand and revenue trends, allowing sales and marketing teams to be ahead of market changes and customer interests.
- AI-driven dashboards bring together performance data from across the teams, providing a bird's-eye view of lead progress, conversion rates, and customer interactions.

AI-Driven Sales and Marketing Alignment Is No Longer Optional

Gone are the days when you could trust in human instinct and a few sales-marketing check-ins. AI makes sure that sales and marketing are seamless data-based operations capable of making immediate decisions that can both increase revenues and keep customers happy.

Businesses that are not scaling AI-powered alignment may be pushed aside in the rush to innovation by faster, tech-enabled competitors. AI is here to stay. The only question is: will your business think differently and embrace the future?

AI in Lead Management and Handoff

One of the primary friction points between sales and marketing is the process of lead qualification and handoff. The challenges are being overcome by AI in the form of predictive analytics, auto-leads scoring, and dynamic content personalisation.

Predictive Lead Scoring Becomes the New Norm

Historically, marketing qualified leads (MQLs) were passed to sales based on rigid criteria and ended up being bad-fit leads that were squandering sales teams' time. AI-driven lead scoring fixes this problem by sifting through historical data, engagement scores, firmographics and behavioural data to predict who is more likely to convert.

Example: Salesforce's AI Einstein Analytics crunches engagement data from multiple touchpoints to pre-qualify leads with predictive scores. This can boost lead conversion by as much as 35%.
https://www.linkedin.com/pulse/35-jump-lead-conversion-rates-salesforce-ai-magic-areyatech-mugsc/

Service Level Agreements (SLAs) in the AI Era

Most companies have SLAs in place to ensure smooth cooperation, defining very structured response times and follow-up procedures. AI improves SLA compliance by automatically deciding on top-priority leads and recommending the best way to engage with a lead at any given time.

Revenue Influence Attribution and AI-Driven Analytics

Attribution of the contribution of marketing activity to revenue has, however, historically been problematic. This has been transformed with AI-powered multi-touch attribution models that serve granular knowledge on customer touchpoint interaction through multiple channels.

Content personalisation and sales enablement with AI

AI-driven content personalisation tools such as **Persado** and **Pathmatics** analyse engagement patterns to recommend content tailored to specific prospect behaviours. By tracking past interactions, AI ensures that every marketing touchpoint adds value.

The Future of AI for Sales and Marketing Alignment

No more are the days of manual lead handoffs, revenue attribution that is out of sync, or one-size-fits-all marketing strategies. AI is not a sidekick; AI is what powers a new era of sales and marketing alignment. Companies that are slow to embrace all that AI has to offer in predictive analytics, automation, and data-driven business intelligence could find themselves at a competitive disadvantage in the B2B space.

Summary:
Sales and marketing alignment is critical for a seamless customer experience. Unified teams improve efficiency, communication, and conversion rates.

- Sales and marketing leaders believe that alignment between their teams drives business growth.
- Traditional alignment relies on manual processes and periodic meetings, leading to inefficiencies.
- AI enables real-time insights, automated workflows, and predictive lead scoring to ensure seamless collaboration.

Case Study: HubSpot's Flywheel Model
HubSpot pioneered the flywheel model, replacing the traditional sales funnel by integrating AI-powered marketing, sales, and service processes. The flywheel approach enables businesses to use AI to continuously learn from customer interactions and refine their engagement strategies.
https://www.hubspot.com/flywheel

5. Account Based Marketing (ABM & ABX) in Action

"ABX isn't just an extension of ABM; it's a fundamental shift. ABX takes the best of ABM, aligns it with customer expectations, and optimises every touchpoint in the buyer's journey".

Jon Miller, Co-Founder of Marketo and Engagio.

For some time now, Account-Based Marketing (ABM) has quickly been hailed as the B2B marketing strategy of the 21st century, as a highly focused strategy which targets only the most potentially profitable key accounts rather than casting a wide net over a broad audience. ABM zeros in and adds real value to relationships with specific companies that closely resemble their Ideal Customer Profile (ICP).

Through personalised messaging, tailored insights, and alignment between sales and marketing teams, ABM is a way for companies to invest more strategically, with companies focusing their resources on prospects who have the most potential to deliver in the long-term. This is what makes it so effective for businesses that are selling in the complicated B2B world, where sales cycles are long, and customers need to be nurtured over weeks, months and even years.

However, as we move into the era of customer experience (CX) and the expanding convolutions of today's B2B buying cycle, the evolution of ABM has become ABX – Account-Based Experience. ABX combines data-driven insights, real-time engagement, and AI-driven automation to ensure that the entire customer experience is as seamless as possible across all customer touchpoints.

ABM has become more successful, leading to the demise of traditional demand marketing, and is game-changing for all revenue-responsible roles;

Forrester highlights that here
https://www.forrester.com/report/the-future-of-account-based-marketing-isnt-scale-its-profitable-growth/RES180931

According to **Gartner**, 77% of B2B buyers describe their latest buying experience as difficult, making it clear that it's time to shift to a more comprehensive, experience-oriented solution.
https://www.forrester.com/report/the-future-of-account-based-marketing-isnt-scale-its-profitable-growth/RES180931

This change is indicative of a bigger trend in the market, one where companies need to transition from lead gen to engagement, retention, and long-term value.

Why ABM Needed to Evolve into ABX

ABM has lately been at the forefront of achieving more engagement and a better return on investment (ROI) when it comes to B2B marketing. However, changing buying behaviours, lengthening sales cycles, and heightening engagement through digital means have made a move to Account-Based Experience (ABX) mandatory. ABX reframes ABM around the buyer's experience. Every touchpoint, from **LinkedIn** messages to analyst webinars, should feel orchestrated, not opportunistic.

The rise of Customer Experience (CX)
B2B buyers demand a B2C-like experience – one that is fully personalised and frictionless. ABX remedies this by being buyer-centric and having consistent engagement on all touchpoints. 88% of customers say the experience a company provides is as important as its products or services, according to **Salesforce**'s 'State of the Connected Customer' Report.
https://www.salesforce.com/news/stories/customer-engagement-research/

The Growing Complexity of B2B Purchasing
The B2B purchase process has grown increasingly complex and includes multiple stakeholders, dozens of digital touchpoints, and extended

consideration periods. ABX is built to cut through this by providing customised, AI-driven engagement at every single stage.

According to **Gartner**, 77% of B2B buyers described their purchase experience as very complex or difficult, and it's easy to see why, given the difficulty of getting many decision-makers and sources of information to march in unison. https://www.gartner.com/en/sales/insights/buyer-enablement

The Shift from Sales-Driven to Buyer-Led

Traditional ABM aligned marketing with sales-oriented outreach. But modern purchasers want to self-educate even before connecting with sales. ABX makes it possible for sales to engage continuously, providing value-based engagement without involving a salesperson every step of the way.

According to **Gartner**, 80% of B2B sales interactions between suppliers and buyers will take place through digital channels, reflecting a buyer-first reality. https://www.gartner.com/en/sales/trends/future-of-sales

ABM and ABX: Why They Matter to B2B Success

The shift from Account-Based Marketing (ABM) to Account Based Experience (ABX) is not just a trend in the industry, it's a strategic move you need to make to succeed in B2B marketing long term. ABX takes the best of what works in ABM and makes customer engagement, retention, and expansion even better. Businesses that succeed in using ABX realise greater customer lifetime value (CLV), better sales and marketing alignment, and better ROI.

The Return on Investment of ABX vs. Traditional B2B Marketing

ABX takes inspiration from Account Based Marketing (ABM) but with an added focus on long-run relationship-building and deep, bespoke-level experiences. It's this change that results in dramatic gains in revenue and marketing efficiency. Companies bring in 208% more revenue from their marketing campaigns.
https://intelligentdemand.com/resources/15-account-based-statistics-that-will-get-you-executive-buy-in/

Eighty-seven percent of B2B marketers said it delivers a higher ROI than

any other type of marketing.
https://insights.strategicabm.com/account-based-marketing-stats

Better sales and marketing alignment

One of the hardest aspects of B2B marketing is the disconnect between sales and marketing teams. How do most traditional demand-gen models work? Marketing produces a large volume of leads (many of which are not SQ) and passes that volume on. ABM and ABX remove this waste by getting sales and marketing on the same page around common objectives and target accounts.

Businesses that have closely aligned sales and marketing departments are 36 per cent more likely to retain customers and 38 per cent more likely to win new customers. Retention matters, as attaining a new customer can cost anywhere from a few times to as much as ten times more than retaining an old one.
https://radiateb2b.com/blog/the-ultimate-page-of-account-based-marketing-statistics/

Deeper Personalisation and More Focused Engagement

One size fits all marketing messages are losing their effectiveness in the B2B world. They want personalised engagements that speak to their specific business issues. ABM and ABX allow companies to deliver account-specific content, messaging and experiences that help potential buyers feel understood and appreciated.

According to **Forrester**, 56% of marketers say personalised content is critical to ABM's success. https://insights.strategicabm.com/24-account-based-marketing-statistics-you-need-to-know-in-2023?

More Opportunities To Retain And Upsell Customers

ABX is not only about acquiring the customer. It elevates the account-based approach to include beyond the first sale, which means ongoing engagement, retention, and expansion. Research shows that firms with a commitment to customer engagement tend to have higher rates of retention and grow quicker within their current account base.

Strengthening relationships and consistently adding value, ABX strategies enable companies to lower churn and grow revenue coming from current accounts.

Shorter Sales Cycles and Larger Deals
With a concentration on high-intent, high-value accounts, ABM and ABX cut down wasted time chasing low-likelihood leads. This enables sales teams to close more deals faster and at bigger average deal sizes.

ABM programs have led to a 20% larger average deal size compared to traditional demand generation programs, according to Radiate B2B. https://radiateb2b.com/blog/the-ultimate-page-of-account-based-marketing-statistics/

ABM and ABX in the Age of Modern B2B Marketing

B2B sales have become more complicated, along with longer sales cycles, multiple decision makers, and increased digital engagement. Old school tactics in marketing are nowhere near equipped to accommodate these changes, so Account-Based Experience (ABX) is the 'future now'. ABX delivers an end-to-end, personalised buying experience from the very first touch, addressing some of the most pressing modern B2B challenges.

Long Buying Cycles Require Sustained Engagement
The B2B buyer's journey is no longer linear. Buyers do a lot of independent research, comparing multiple vendors and engaging in multiple conversations with stakeholders before they even make a purchase.

This is what ABX addresses by delivering always-on, informed engagement with personalised content, predictive lead nurturing and data-driven sales interactions.

Multiple Stakeholders are Involved in Decision-Making
The average B2B buying group now includes six to ten decision-makers, each with different priorities and concerns. Generic marketing efforts fail to

address these diverse needs effectively.

ABX solves this by leveraging AI-based segmentation and intent data to tailor messaging and content to the unique issues of individual buyers. This method ensures that sales and marketing messages ring true for the technical buyers, as well as the executive managers who hold the purse strings.

Buyers Demand Highly Personalised Experiences

B2B buyers have begun to expect the same kind of personalisation they get on the consumer side. Those that don't deliver on this lose engagement and trust.

Seventy-five percent of B2B buyers don't want to talk to a salesperson at all, relying instead on search and online content to make their decision. https://www.demandgenreport.com/uncategorized/scale-smarter-not-harder-adapting-your-b2b-buyer-journey-in-2025/48793/

Real-time behavioural insights, machine learning-driven recommendations, adaptive content and delivery strategies make personalisation at scale possible with ABX.

The Need for Multi-Channel Engagement

B2B buyers are interacting across a multitude of channels, including social media, industry forums, and direct website visits. A piecemeal marketing strategy will produce inconsistent communications and wasted opportunities.

ABX orchestrates a unified, omnichannel strategy, ensuring that interactions across email, social media, content platforms and direct sales engagements are aligned and data-driven.

Implementing ABM and ABX Strategies: Best Practices

The move from traditional marketing methods to Account-Based Marketing (ABM) and Account-Based Experience (ABX) represents a transformative new phase in B2B marketing that is centred around delivering personalised and friction-free experiences across the customer journey. In order to run ABX effectively and keep up with the hottest trends, companies need to adopt a strategic, data-driven approach that prioritises customer engagement and retention.

Align Cross-Functional Teams

And ABX is only effective if marketing, sales, and customer success are closely aligned. By aligning these teams, you ensure that there is one concerted effort behind account engagement, which in turn improves the total customer journey.

- Set shared goals: Set collective goals and KPIs that encourage collaboration and responsibility across departments. https://www.folloze.com/blog/article/abx-guide
- Foster Consistent Communication: Conduct cross-functional briefings on account strategies, best practices, and challenges.

Leverage Data-Driven Insights

Data is the cornerstone of ABX, delivering rich and accurate data that allows businesses to craft ultra-personalised and highly targeted experiences for target accounts.

- Leverage Intent Data: Use a specific set of tools that track and analyse online behaviour to spot accounts that are showing purchase intent and engage with them when it matters. https://www.revsure.ai/blog/abx-strategies-for-b2b-marketers-in-2025
- Use advanced analytics: Track engagement at each touchpoint to uncover preferences and purchase behaviour, resulting in more personalised interactions.

Create Personalised Content and Messaging

Content that speaks to the personal needs and issues facing target accounts is the key to ABX's success.

- Create Account Relevant Content: Create personalised content like case studies, industry reports, or personalised video messages that speak directly to the issues an account cares about. https://www.heinzmarketing.com/blog/keys-to-successful-abx-rollout/
- Tailored Communication: Tailor messaging to reflect account-specific priorities for increased relevance and engagement.

Implement Multi-Channel Engagement Strategies

B2B buyers engage across multiple platforms, from social media to direct website interactions. A fragmented marketing approach risks inconsistency and missed opportunities.

- Embrace Omnichannel: Synchronise engagement across email, social media, webinars, direct mail and industry events to ensure a consistent experience.
- Analyse Channel Performance: Keep up with what platforms are performing best and adapt accordingly.

Embrace AI and Automation

AI and automation make ABX more efficient and scalable through personalisation. These solutions enable companies to anticipate rather than react to customer needs with real-time engagement.

- Predictive Analytics: AI will predict how accounts will behave so that they can proactively personalise your marketing and sales touchpoints.
- Automatable processes: Cut down on menial tasks like lead scoring, follow-ups, and reporting - so your team can concentrate on high-value engagements.

Prioritise Customer Experience and Retention

ABX is more than just acquisition; it's also about long-term engagement, satisfaction, and customer loyalty. Companies that prioritise customer success see higher retention and advocacy levels.

- Customer Success Programs: Create programs that enable customers to realise the full potential of your product or service, securing sustained relationships.
- Feedback Loops: Develop and maintain a system of feedback collection and analysis, driving continuous improvement.

Measure and Improve Performance

Continued evaluation and evolution of ABX strategies are important for remaining influential and supporting growth.

- Create Meaningful Metrics: Monitor engagement rates, conversion rates, and CLV to measure success.
- Performance Reviews: Review how you are doing, create opportunities

for improvement and chart your improvement.

Implementing ABX successfully requires a strategic approach that aligns teams, harnesses data, embraces technology and prioritises customer experience. By adopting best practices and anticipating future trends, businesses can drive stronger engagement, retention, and growth in an increasingly competitive B2B landscape.

ABM & ABX Tactics for Maximum Engagement

Effective ABM and ABX strategy entails more than just identifying target accounts, it demands a structured engagement plan that delivers meaningful interactions across multiple touchpoints. In ABM, the goal is to convert high-value accounts, while in ABX, the objective extends further to delivering long-term value and deepening relationships with customers.

To be ahead of the curve, businesses need to be strategic, data-oriented, and focus on customer engagement and retention.

Align Cross-Functional Teams

Effective ABX needs good marketing, sales, and alignment with customer success. By aligning these functions, you will have a consistent approach to account engagement, as well as a unified customer experience.

- Define Common Objectives and KPIs that promote collaboration and accountability across Departments.
- Demand Collaboration: Schedule cross-functional meetings to talk about account strategies, share insights, and resolve obstacles together.

Leverage Data-Driven Insights

Data is fundamental to ABX, enabling organisations to build a specific and personalised engagement for key accounts.

- Use Intent Data: Implement tools that analyse online behaviour to identify accounts showing signs of purchase intent, allowing for timely and relevant engagement. https://www.revsure.ai/blog/abx-strategies-for-b2b-marketers-in-2025
- Apply Advanced Analytics: Monitor engagement across multiple touchpoints to gain insight into preferences and buying behaviours, ensuring personalised interactions.

Develop Personalised Content and Messaging

Creating content that resonates with the specific needs and challenges of target accounts is essential for ABX's success.

- Produce Account-Specific Content: Develop tailored materials such as case studies, industry reports, and personalised video messages that speak directly to an account's concerns.
- Customise Communication: Adapt messaging to align with the unique priorities of each account, improving relevance and engagement.

Roll Out Multi-Channel Engagement Strategies

B2B buyers engage across multiple platforms, from social media to direct website engagement. A disconnected marketing strategy can lead to a lack of cohesion and missed opportunities.

- Adopt an Omnichannel Approach: Coordinate engagement across email, social media, webinars, direct mail, and industry events to create a seamless experience.
- Monitor Channel Performance: Regularly assess which platforms drive the most engagement and adjust strategies accordingly.

Adopt AI and Automation

AI and automation make ABX more efficient and allow it to scale through personalisation. These tools allow companies to be proactive in addressing a customer rather than reactive.

- Predictive Analytics: Leverage AI to predict actions and independently optimise marketing and sales touchpoints.
- Automation: Reduce unnecessary repetition of tasks, such as lead scoring, follow-ups, and reporting, so teams can focus on higher-value interactions.

The Future of ABX: Emerging Trends and Strategic Considerations

As B2B marketing continues to evolve, Account-Based Experience (ABX) is becoming increasingly sophisticated. With advancements in AI, predictive analytics, and intent-based marketing, businesses must stay ahead of the curve to maintain competitive advantage. The future of ABX is defined by hyper-personalisation, automation, and seamless omnichannel experiences.

Predictive Analytics

AI and ML are revolutionising ABX with actionable real-time data insights that enable enterprises to predict customer needs and behaviour.

- With predictive lead scoring, marketers and salespeople alike can prioritise which accounts to go after.
- Content recommendations are powered by AI to make sure target accounts get the message that matters at the right time.
- According to **McKinsey**, with Artificial Intelligence driven personalisation, marketing effectiveness can be improved by 30%.

https://www.mckinsey.com/featured-insights/mckinsey-explainers/what-is-personalization

The Rise of Intent-Based Marketing

Intent data is a critical component of ABX, as it pinpoints accounts that are actively researching solutions in real-time. This enables marketing teams to connect with potential clients when they are most interested.

- Third-party intent data vendors like **Bombora** and G2 provide more detailed insights on buyer intent.
- Real-time engagement triggers allow for more proactive engagement.
- **Forrester** discovered that with the assistance of **6sense**, companies leveraging intent data are able to drive a 4X increase in conversion rates. https://6sense.com/resources/pdfs/forrester-tei-report-2022#

Hyper-Personalisation at Scale

Personalisation has always been key to ABX, but new artificial intelligence (AI)-based tools mean that highly customised experiences can be delivered at scale.

- AI-driven chatbots and virtual assistants provide real-time, contextual engagement.
- Dynamically personalised content adapts based on user interactions.
- **Salesforce**'s 'State of the Connected Customer' report found that 66% of B2B buyers expect brands to understand their unique needs. https://www.salesforce.com/resources/research-reports/state-of-the-connected-customer/

Omnichannel Engagement Will Be Key to ABX Success

B2B buyers interact with brands in multiple places, and great ABX initiatives need to be able to present a unified front in both digital and offline environments.

- Social Selling will be more prevalent, and **LinkedIn** is set to dominate B2B interaction.
- Real-time buyer engagements will be improved even further with conversational marketing tools like **Salesloft (Drift)** and **Intercom.**

The Role of ABX Expands Beyond Acquisition to Retention and Expansion

ABM is usually centred around getting new accounts. ABX drives value over time by addressing both retention and growth within an account.

Key Developments:

- AI-powered customer success platforms help identify accounts at risk of churn.
- Upsell and cross-sell strategies are optimised using predictive data analytics.
- A study revealed that 80% of a company's future revenue will come from 20% of its existing customers.

https://www.bigcommerce.com/glossary/how-retain-customers-and-spend-less/

The transition to customer-first, AI-driven marketing is changing the way companies target key accounts. ABX is more than just a marketing tactic; it's a way of thinking about the relationship between sales, marketing, and customer success in pursuit of lasting business growth.

ABX Is Not a Marketing Strategy – It's a Business Growth Strategy

ABX isn't just the future of Account-Based Marketing (ABM), it's the new way of doing business that aligns marketing, sales, and customer success teams to deliver a seamless experience for your high-value accounts.

- Organisations that prioritise sales and marketing alignment are nearly three times more likely to exceed customer acquisition targets.

https://www.linkedin.com/posts/gartner_how-to-build-sales-marketing-alignment-activity-7217154736913817600-h4b2/

- 91% of companies using ABM increase their average deal size, with 25% reporting an increase of over 50% https://www.hartehanks.com/blog/measuring-success-the-5-rs-of-account-based-marketing-metrics/

These insights demonstrate that ABX, as an extension of ABM, drives not only sales but also long-term business growth.

Competitive Differentiator will be Customer Experience and Retention. With the B2B sector more crowded than ever, businesses that focus on the experience of customers in order to retain them will be ahead of the rest. Using ABX, companies can shift from specialising in acquisition and lead generation to driving loyalty, customer lifetime value and long-term relationships.

The Integration of ABX into the Wider Marketing Ecosystem

ABX does not exist in isolation; it must be seamlessly integrated with broader marketing strategies, including content marketing, social selling, and multi-channel engagement.

- B2B buyers interact with an average of 10+ marketing touchpoints before making a purchase.
  https://huble.com/blog/account-based-marketing-statistics

This reinforces the need for ABX to be fully integrated with content, thought leadership, and digital engagement strategies.

To see how **Demandbase** are viewing this download their guide to GTM with account intelligence and ABM/ABX. https://www.demandbase.com/se-unspam-your-brand-account-based-marketing/

Summary:

ABX is no longer an experimental approach; it is becoming the industry standard for modern B2B engagement. Companies that fail to adopt ABX will struggle to keep up with the increasing demands for personalisation, customer experience, and AI-driven engagement.

- ABX is a strategic business framework, not just a marketing tactic.
- AI, automation, and data-driven insights are shaping the future of ABX.
- Customer experience and retention are becoming the most valuable growth levers.
- ABX must be integrated into the wider B2B marketing and sales ecosystem.

Case Study: Snowflake's Hyper-Personalised ABX Approach

Snowflake's strategic implementation of ABX, characterised by hyper-personalisation and cross-functional collaboration, resulted in significant improvements in account engagement and sales outcomes. This case underscores the potential of ABX to drive business growth by delivering tailored experiences that resonate with high-value accounts.

https://green-hat.com.au/articles/great-abm-case-studies-worth-knowing-about

6. Thinking Long Term: The Strategic Value

"People don't buy what you do; they buy why you do it."

Simon Sinek, Leadership Expert.

The Importance of a Long-Term Perspective in B2B Marketing

For B2B marketers, the temptation to focus on quick wins through short-term tactics is often strong, especially with pressures related to quarterly targets, budget limitations, or stakeholder expectations. But the short-term gains never pave the way for the future. Rather, for B2B marketing, the long-term perspective is essential when it comes to creating brand equity, developing deep customer relationships and establishing a business as an industry leader.

A balanced approach of long-term and short-term is typically optimal. In their book The Long and Short of It, marketing scholars Les Binet and Peter Field note that firms that spend at least 40% of their marketing budgets on brand-building have better customer retention and lower customer acquisition costs over time. This implies that as important as it is to have a long-term plan, it should be combined with effort-driven short-term ones that help keep your business on a roll.
https://www.amazon.co.uk/Long-Short-Balancing-Long-Term-Strategies/dp/085294134X/

The B2B sales process is long and complicated; it touches several people in an organisation, and obviously, it has high risks and high investments. Align your marketing in line with this long buying cycle and you'll have a consistently growing trajectory. B2B brands need to invest in long-term brand building because it targets 'out of market customers'. The B2B

Institute, 2030 B2B Trends:
https://business.linkedin.com/content/dam/me/business/en-us/marketing-solutions/case-studies/images/namer-pdf/lin-b2b-2030-trends-pages-v4-new.pdf

Studies have shown that 95% of B2B prospects are not actively looking to buy at any given moment, which is why you need to maintain ongoing engagement and a brand presence so that these companies remember you when they are. It's a bitter pill to swallow, but for the majority of your future purchasers, you are a stranger. That's why brand matters.
https://marketingscience.info/ehrenberg-bass-95-of-b2b-buyers-are-not-in-the-market-for-your-products/

The Long Game in Marketing: ESG Factors

Sustainability and ethical marketing are standard in B2B practice. Sustainability and ethical marketing are no longer nice in a B2B strategy. Today, companies are facing growing demands to become more ESG-compliant. These values are becoming new vectors for trust, loyalty and differentiation. Deloitte research shows that 75% of B2B buyers favour providers with explicit ESG commitments over those without, pointing to a market where procurement is increasingly about more than price and performance.
https://deloitte.wsj.com/sustainable-business/looking-to-improve-b2b-customer-trust-demonstrating-sustainability-can-help-ddeeb940

However, it is important that ESG initiatives are genuine and actionable rather than pure marketing exercises. Many corporations come under fire for greenwashing, green marketing that is spin, exaggeration or falsely claiming sustainability efforts. In an age of examination of social activism, your ESG position isn't just a policy; it's your character on show. Transparency in communication, ethical treatment of data and measurable ESG commitments add credibility and are aligned with the ethos of current generation decision-makers.

When companies incorporate ESG into their long-term marketing plans,

they are not only future-proofing their strategy, but they are also developing more profound emotional relationships with their stakeholders. For instance, **Unilever**'s Sustainable Living Plan has been a cornerstone of its marketing strategy for the long-term and it is evident how sustainability values result in very real brand equity and consumer support.
https://www.unilever.com/sustainability/responsible-business/sustainable-development-goals/

Benefits of Long-Term Marketing Strategy in B2B Businesses

A solid, effective B2B marketing plan for the long-term produces several advantages that also help your business thrive. Short-term wins might bring in quick and easy revenue, but you need long-term strategies that will help build a powerful brand, create trust and plan for predictable growth.

Developing Brand Trust and Loyalty:

Trust is key in B2B, as decisions may involve high amounts of money or huge operational changes. In B2B, trust is everything. The truth, in time, will become your greatest risk reducer and strongest sales rep.

Nearly two-thirds of consumers now purchase or recommend brands based on their beliefs, according to the Edelman Trust Barometer.
https://www.edelman.com/trust/2022-trust-barometer

Staying Ahead of Competitors:

A forward-thinking approach enables brands to position themselves as thought leaders, keeping potential customers on the hook. This proactive strategy can set a company apart from competitors who may be too concentrated on short-term sales methods.

Effective thought leadership is more powerful than traditional methods, making people more inclined to find you and even pay a premium for your expertise. As found in the 2024 Edelman-LinkedIn B2B Thought Leadership Impact Report.

https://www.edelman.com/expertise/Business-Marketing/2024-b2b-thought-leadership-report

Driving Predictable Growth:

A good long-term strategy doesn't necessarily feed leads directly to the sales team but rather feeds the sales team leads that have been nurtured over time and are ready to buy. This stability reduces the dependence on reactive marketing, which tends to be inconsistent and volatile.

Diversity and Inclusion in B2B Marketing

Diversity and inclusion have become business imperatives now that help you have long-lasting success with B2B marketing, not just feel-good social initiatives. The benefits of inclusive marketing extend beyond building stronger customer relationships to creating competitor advantage in a globalised marketplace.

The Business Case for B2B Marketing Diversity:

A **McKinsey** study found that companies in the top quartile for ethnic and cultural diversity are 36 percent more profitable. Meanwhile, companies with more gender diversity are 25% more likely to outperform their industry. https://www.mckinsey.com/featured-insights/diversity-and-inclusion/diversity-wins-how-inclusion-matters

What does this mean for B2B marketing?

- We have a more diverse group of decision-makers, so marketing should represent the perspectives of a wide range of stakeholders.
- Through inclusive marketing, you grow market share and improve your brand image.
- Companies with diverse suppliers are designed to bring resilience and innovation into the value chain.

Key Methods for Embracing Inclusive B2B Marketing

Diversity of Representation: In Images, Language, and Messaging

- Above all, inclusive imagery should be used that is representative of various ethnicities, genders, ages, and abilities.
- Use gender-neutral and culturally respectful language in marketing materials.

Constructing Campaigns That Resonate with Underrepresented Groups

- Showcase case studies and testimonials from all the different voices in the industry.
- Create thought leadership content that tackles challenges facing underserved communities in B2B.

Working with Inclusive Organisations and Suppliers

- Work with minority-owned, women-led, and disability-inclusive businesses to reinforce supplier diversity.
- Get behind industry initiatives that are driving diversity, such as The Valuable 500, (https://www.thevaluable500.com/) a global campaign that is demanding disability inclusion in business.

Microsoft has built-in diversity and inclusion within its B2B marketing by ensuring that the advertising mirrors true diversity. The company has also pledged to increase its spend with diverse suppliers by $150 million. https://www.microsoft.com/en-us/racial-equity-initiative

The Elements of a Successful Long-Term B2B Marketing Plan

A successful long-term B2B marketing plan is complex, involving elements of brand development, customer insights, on-message communications and data-driven adaptation. In my experience, the marketers who defend the long-term value are often the ones who get the board positions. Here are some of the main components of a successful long-term strategy.

Customer-Centric Approach

The essence of a successful B2B marketing plan is finding out customer requirements. Businesses that centre their marketing strategy on addressing customer pain points will create lasting relationships and establish themselves as trusted partners.

An analysis by **Gartner** discovered that 86% of B2B buyers would like companies to understand their needs and make customer experience more

proactive, while 80% of organisations that personalise interactions see a lift in retention.
https://isbm.org/business-readings/gartner-survey-reveals-86-of-b2b-customers-expect-companies-to-be-well-informed-about-their-personal-information-during-service-interactions/

Developing a Clear and Evolving Value Proposition

It's what differentiates your business from your competitors and why your prospects should choose you. This value proposition can change over time in response to shifts in customer needs or market conditions, but its essence should be constant.

A **HubSpot** article states that companies with well-aligned sales and marketing organisations experience 24% faster three-year revenue growth and 27% faster three-year profit growth, highlighting the need for clear and ongoing value prop work.
https://blog.hubspot.com/sales/stats-that-prove-the-power-of-smarketing-slideshare

Brand Consistent Messaging Across All Channels

Repetitive messaging is a trust-and-recognition imprint. When a brand maintains a consistent voice, design ethos and values from one touchpoint to the next (social media to email to website, etc), that consistency adds levels of professionalism and trust.

According to Lucidpress research, reliable branding can add a third (33%) to the bottom line. https://www.prnewswire.com/news-releases/study-finds-companies-with-consistent-branding-can-see-up-to-33-increase-in-revenue-300967219.html

Content Strategy & Thought Leadership

It's no secret that in B2B marketing, thought leadership is possibly the single most powerful trust-building tool. Companies can establish themselves as authority voices within their industry, providing valuable insights and sector knowledge.

According to the **LinkedIn**-Edelman Thought Leadership Report, 64% of B2B buyers say that thought leadership content has a major impact on their purchasing decisions.
https://www.edelman.com/expertise/Business-Marketing/2024-b2b-thought-leadership-report

Data-Driven Decision Making

Data-driven insights should be the foundation of a sustained marketing effort that is iterated. By using customer behaviour data, market trends and campaign performance, marketers can make adjustments on the fly while keeping in line with the overall objectives.

In fact, **Forbes** reports that data-driven organisations are 23 times more likely to outperform their competitors in customer acquisition. https://www.forbes.com/councils/forbeshumanresourcescouncil/2023/07/18/being-data-driven-is-likely-your-best-bet/

Alignment with Sales Goals

In a long-term marketing strategy, goals must directly correspond with sales objectives so that both teams work in the same direction. And businesses that have sales and marketing aligned sell more and have better customers.

According to a **HubSpot** report, organisations with strong sales and marketing alignment see a 38% higher sales win rate.
https://blog.hubspot.com/sales/stats-that-prove-the-power-of-smarketing-slideshare

Creating Buyer Personas and Mapping the Customer Journey

In a sense, it is no more than what a strategic approach to marketing demands: knowing your customers. Personas and journey mapping enable you to create marketing messages that 'speak' to your ideal clients. I have, however, seen them abused and applied extra segmentation where none was needed, so use them judiciously.

Creating Buyer Personas

Buyer personas are fictional, generalised representations of your ideal customers, formulated based on market research and real data about your existing customers. Some details each persona should have:

- Title and role in decision making
- Company size and industry
- Challenges and what sucked.
- Goals and objectives
- Preferred means of communication

Mapping the Customer Journey

Once the personas are identified, charting their path to purchase, starting from the awareness stage, helps marketers get insight into decision-making and identify important engagement touchpoints.

Customer Journey Stages:

- Awareness Stage – The buyer is aware that they have a problem.
- Consideration Stage – The prospect is looking at various options.
- Decision Stage – The buyer is prepared to select a solution.

A **McKinsey** review discovered that B2B shoppers consume more than 10 content pieces while on their path to purchase, illustrating the significance of a well-mapped journey.
https://www.mckinsey.com/capabilities/growth-marketing-and-sales/our-insights/the-new-b2b-growth-equation

Understanding the problems and gaps in each stage

Various stages of the customer journey imply different challenges and information demands. By addressing these pain points with relevant content, prospects are properly funnelled into a purchase.

Using AI and Machine Learning to drive Personalisation

AI-powered tools can improve customer journey mapping in the following ways:

- Monitoring customer behaviour to see where they are in the buying cycle.
- Recommending user-specific content by using interaction history.

- Automating email sequences as trigger chains with targeting, delivering content based on the right stage.

According to an ON24 report, 88% of companies are looking to use AI to strengthen their personalisation strategies, making it essential for future marketing spend.
https://www.on24.com/blog/6-striking-statistics-that-show-the-power-of-personalization-infographic/

Developing Your Long-Term Content Marketing Plan

Content marketing is important for long-term B2B value. Content is not like a short, punchy ad, but something that continues to work for you, it nurtures leads and builds authority over time. When it comes to both the creative and commercial aspects of your business, a solid content strategy ensures that your brand remains front and centre in the hearts, minds and wallets of its customers.

Phases of Content in a Long-Range Plan

An effective content strategy will speak to the buyer at every stage of their journey, including:

The Awareness Stage - Educational Content

At this point, leads are conducting industry research and exploring pain points. Content should be informative, not pushy.

- Blog articles about industry challenges
- Research and white papers
- Industry reports
- Educational videos
- Infographics explaining key concepts

Consideration Stage – In-depth Content

Once potential buyers recognise what the problem is, they start looking for ways to solve it. This is where comparative and value-led content comes in:

- White papers
- Webinars with industry experts
- Product comparison guides

- Real-world results from case studies

Decision Stage - Content to persuade the audience

Prospects at this stage are making a final decision. Content would need to be focused on ROI, credibility and reassurance:

- Demos, ROI calculators, testimonials, free trials.
- Free trials and demos
- ROI calculators
- Testimonials and client success stories

Consistent Publishing and Distribution

A well-executed content strategy relies on consistency:

- Plan blogs, videos and emails via a content calendar.
- Re-purpose content in different formats (such as turning a webinar into a white paper).
- Use all available distribution channels (social media, email, industry events).

According to a Content Marketing Institute study, companies that publish content consistently receive 7.8x more website traffic than those that do not. https://contentmarketinginstitute.com/articles/content-marketing-stats/

Investing in Brand Building

The reason why this missing link is overlooked in B2B is that we focus more on product features than on emotional engagement. However, a good brand builds loyalty, lowers perceived risk, and simplifies sales discussions.

Factors of Brand Building

Brand Consistency

The power of visual branding, messaging and tone is key to recognition and trust across all channels.

Lead The Thought And Expertise

Thought leadership sets your brand up as an authority. Writing authoritative content, speaking at events and participating on social media shows off your industry knowledge.

Connection Through Storytelling
Storytelling brings your brand to life, giving it an emotional angle to connect to. Use case studies, testimonials, and success stories to demonstrate the real-life benefits of your products.

Measuring Success for the Long-Term

A long-term plan requires patience, as results may not be immediate. By following the right metrics, you stay on the right track, and you measure your success by brand power, customer loyalty, and solid sales growth.

Key Long-Term Metrics

Brand Awareness and Perception: Measure brand sentiment and share of voice, brand recall from surveys and social media listening.

Customer Lifetime Value (CLTV): CLTV is the total amount of revenue you anticipate a customer will spend over the entire relationship that they maintain with your company. A higher CLTV translates to stronger loyalty.

Content Engagement Metrics: Page views, social shares and average time on site indicate how well your content is resonating.

Quality of Leads and Conversion Ratio: Track efforts to build trust and how they can lead to the right leads and conversions.

Pipeline Development: The sign of a healthy and growing business is an increased pipeline of high-quality prospects.

Adapting to Market Changes

A long-range strategy does not mean rigidity. The B2B environment is changing at a fast pace, and you need to be flexible to stay relevant. To develop adaptability, you need:

Periodic Strategy Review: Organise periodic reviews of the strategy in each quarter or year.

Follow Industry Trends: Keep an eye on new trends and customer requests to keep ahead of the game.

Be Flexible in Execution: Change tactics but keep the long-term vision intact.

How to Use AI and Machine Learning for the Long Term

The rise of AI and machine learning in B2B marketing, which are transforming the way companies interact, evolve and expand.

AI in Customer Journey Mapping and Personalisation:

Real-Time Customer Journey Mapping: AI-based tools track touchpoints and analyse customer behaviours in real-time to trace the most granular buyer journeys.

Hyper-Personalised Content: AI driver personalisation learns user preferences and ensures most relevant content is served up.

Machine Learning for Predictive Analytics and Trend Forecasting

Predictive Lead Scoring: Machine learning models analyse vast datasets to identify high-value leads.

Trend Forecasting and Market Shifts: AI predicts future market trends by analysing industry data, customer behaviour, and competitor activities.

Adaptive Learning Using Customer Data

Establish Feedback Loops for Iterative Improvement: Feedback loops, made possible by AI, are used to analyse customer interactions and continually improve marketing strategies.

Real-Time Personalisation Adjustments: Machine learning fine-tunes personalisation strategies according to response trends as they occur.

Tactical Adoption of AI and Machine Learning

Invest in a Scalable Data Infrastructure: AI and machine learning are only as good as their underlying data, so you need to invest in a cloud-based CRM or a data warehouse.

Define Goals and Objectives: Clearly define your goals (such as increasing lead to conversion rate or decreasing churn).

Pilot AI in small, controlled areas: Demonstrate AI efficacy through limited pilots and use cases (e.g., automated lead scoring).

Continuous Model Review and Tuning: AI models must be constantly adjusted to account for customer behaviour that changes.

Educate Your Team in AI Tools and Insights: Sales and marketing teams need to know and trust AI-made insights.

Summary:
Long-term thinking isn't just a strategy in B2B marketing, it's a mindset. It helps to build a kind of sustainability, trust, and resilience. Marketing based on ESG plays the long game and leads to enduring value.

- Build sustainability into your marketing.
- Concentrate on establishing trust and credibility, not on the quick win.
- Highlight your company's dedication to ESG values.

Case Study: Unilever's Sustainability Goals

Unilever's commitment to sustainability influenced its long-term marketing strategies. By integrating environmental and social governance (ESG) principles, Unilever built trust with partners and customers. Initiatives like "Sustainable Living Plan" showcased a transparent and ethical approach, attracting like-minded B2B collaborators.
https://www.unilever.com/sustainability/responsible-business/sustainable-development-goals/

7: Content Strategy and Thought Leadership

"Content marketing is not about what you sell; it's about what you stand for."

Joe Pulizzi, Content Marketing Expert.

B2B marketing has evolved beyond traditional sales-driven approaches. Companies who leverage rich content to become recognised leaders in relevant topics don't just increase their credibility; they increase overall demand and build trust while also encouraging buyers to say "yes" at all stages of the buyer's journey. In order to succeed in a digital-first world where AI-powered tools set the standard for consumer expectations, content strategy is not simply just a nice-to-have form of marketing; it is rather an essential driver for business success.

Craig Davis, Chief Creative Officer, Worldwide at JWT, once said "We need to stop interrupting what people are interested in and be what people are interested in." This quote captures the transition from outbound to inbound marketing that we've witnessed over the past few years, particularly the rise of content's importance.

This is especially true for B2B, where purchasing decisions typically are complex and involve many decision makers. Good, data-driven content allows brands to explain their value prop, educate potential customers, and drive brand preference.

The demand for content personalisation, automation and predictive analytics has been driven even faster by the rise of AI. Organisations that don't implement AI-powered content strategies will be at a disadvantage in trying to capture the attention of B2B audiences that are becoming more and more sophisticated and expect personalised, data-driven insights.

Creating a Strong Content Strategy

If you're hoping to grow your B2B business, and you aren't using a B2B content strategy, I predict this year will illuminate the necessity. Your content strategy must be a methodical, data-informed strategy that is conducive to business goals, targets audience receptivity and uses AI-driven insights for optimisation.

A truly great content strategy rests on five important pillars: Knowing the audience, Planning the content, Executing and publishing the content, Distributing the content, and Measuring how it performs. All of these components serve to make content efforts more purposeful, measurable, and dynamic in response to evolving industry trends.

Knowing the Audience and Their Content Needs

Understanding your audience is the first step to successful content marketing. Unlike customers in the B2C market, the B2B buying group is frequently a decision-making unit with many stakeholders, each of whom has unique information requirements. With AI and machine learning tools, marketers are able to assess user behaviour, track content consumption flows, and segment audiences with intent data.

Building Out The Content Plan

Planning your content on a calendar ensures that proper topics, formats and publishing timings work with your marketing goals. AI-driven tools can now predict content; they know what hot topics are and what's trending, online and otherwise.

According to **HubSpot**, more than 41% of marketers measure the effectiveness of their content marketing strategy in terms of sales.
https://www.hubspot.com/marketing-statistics

An effective plan includes content formats tailored to each stage of the buyer's journey, such as educational blogs for awareness, case studies for consideration, and product comparisons for decision-making.

Content Delivery and AI-Optimisation

The integration of AI into content execution is changing the way B2B companies produce, disseminate and perfect their content. AI tools such as natural language generation, predictive analytics and automated content personalisation are increasing engagement by providing highly relevant messaging at scale.

According to a study by **McKinsey**, companies that leverage AI-powered content personalisation achieved 10 to 15 percent higher engagement and conversion rates.
https://www.mckinsey.com/featured-insights/mckinsey-explainers/what-is-personalization

Marketers are increasingly using AI to create hyper-personalised content faster and more efficiently – 66 percent, according to **Salesforce**.
https://www.salesforce.com/resources/research-reports/state-of-marketing/

The execution of the content also needs to be nimble; it should evolve following engagement data, SEO performance, and user feedback loops.

Measuring Content Performance and Ongoing Optimisation

One constant in successful content strategies is data-driven iteration. Real-time AI-driven analytics enables enterprises to measure engagement KPIs, conversion rates and content ROI.

According to a study from Porch Group Media, companies that utilise marketing attribution models witness a 10-30% improvement in marketing effectiveness.
https://porchgroupmedia.com/blog/marketing-attribution/

Using AI-powered sentiment analysis, A/B testing, and behaviour tracking, companies can continuously improve their content formats, distribution strategies, and messaging to get the best results.

This is How AI is Revolutionising B2B Content Strategy

AI is not just an add-on in content marketing anymore. It is a fundamental driver of efficiency, personalisation, and strategic decision-making. If your brand doesn't already have AI-driven analytics, predictive content modelling and automation on its roadmap, you're falling behind competitors who are adopting data-led marketing innovation.

In a time when content can't just 'inform', content needs to be strategic, measurable, and constantly optimised. B2B companies need to rethink the way they do content marketing.

Increased Uptake of AI in Content Format Selection

AI is helping marketers understand which content formats perform best with predictive analytics and user intent data informing strategy. Machine-learning models are able to evaluate performance history and recommend optimised structures of content for different kinds of audience components.

According to another research by **Salesforce**, 52% of marketers are using AI to automate distribution and format optimising content, which helps them market in a more targeted manner. https://www.salesforce.com/resources/research-reports/state-of-marketing/

As the use of AI grows, companies that place their bets on automated content intelligence will be able to win competitively by producing personalised, high-performing content formats at scale.

The Value of Creativity in Content Creation

Creativity in B2B content is something that's frequently neglected, yet so essential for desirability, recollection, and impact.

Creative Formats and Visuals

Incorporating infographics, interactive tools, and video into your content strategy makes it much easier to share complex messages quickly. For example, interactive calculators or assessment tools make it possible to receive instant analysis, and that can be the type of thing that gets decision makers' attention.

Storytelling Techniques

Storytelling makes your brand human, giving form and character to the world of abstractions. A good narrative within a case study or testimonial can show the real-life uses and concrete benefits of your solution.

This Harvard Business Review article explains the neuroscience of storytelling: https://hbr.org/2014/10/why-your-brain-loves-good-storytelling

Experimenting with Tone and Style

Traditional B2B content is typically serious and formal, but loosening it up a bit can help encourage engagement, particularly on social media. Trying different tones and styles enables a brand's personality to come through, and, in turn, the content you produce feels less corporate and more human.

Thought Leadership: An Asset for Business Growth

Thought leadership is crucial to B2B marketing strategy because it presents a company as an authority in a certain field or industry. Thought leadership content needs to be more than just selling; it should offer knowledge that benefits your audience, i.e. expert advice and original studies that establish trust with prospects. Is your thought leadership content original? Citable? Modular? Indexed? Quoted by others? If not, it's not thought leadership, it's just content.

According to a report from Edelman-**LinkedIn**, over 75% of decision-makers and execs say that a nugget of thought leadership had prompted them to further investigate a product or service that they hadn't been considering before. Also, 9 in 10 B2B decision-makers and C-suite executives are more willing to engage with a vendor that consistently produces high-quality thought leadership.

https://www.edelman.com/sites/g/files/aatuss191/files/2025-01/2025%20Edelman%20Trust%20Barometer%20Global%20Report_01.23.25.pdf

Some of the characteristics of successful thought leadership:

- Writing long form reports and research papers to create original, data-driven insights.
- As a contributor resource for blogs, podcasts & speaking engagements on industry trends.
- Webinars and panel discussions are included to impart knowledge and network with the industry.
- Using AI to generate insights-driven content and automated research summaries that brands a company as a thought leader.

Companies that consistently invest in thought leadership content get rewarded with greater levels of trust in their brands, stronger customer engagement, and greater customer loyalty.

How Thought Leadership And AI-Powered Content Formats Are Driving Business Growth In B2B

B2B content marketing is evolving as artificial intelligence comes of age to determine what kinds of content and thought leadership are most effective. Companies should concentrate on creating content that is of high value and creates education, logic and trust, with the use of AI insights for effectiveness. The ones who think differently, who are pro-automation, and who lead with expertise will define how we measure the industry and facilitate real engagement in the AI era.

Methods of Distributing/publishing and promoting your content

Good content is only half of a strong B2B marketing plan. Equally important is getting the content in front of the appropriate audience through the correct channels. Content distribution approaches need to be multi-channel, data-driven, and AI-empowered to optimise engagement and ROI.

A stand out example of content as a long-term trust platform is **Salesforce+**, the company's owned media streaming service. Rather than publish scattered content across isolated channels, **Salesforce+** brings customer stories, product deep dives, and C-suite interviews into a single narrative environment. It treats thought leadership not as a campaign but as programming, blending education with brand voice at scale. The result is not just increased brand affinity but greater message consistency and deeper relationship-building across buyer types. https://www.salesforce.com/plus

As digital marketing competition heats up, companies must start to think about content distribution in a new way. Artificial intelligence driven insights, automation and predictive analytics can increase reach and at the same time optimise the mode and format for potential content delivery.

Maximum Reach With Multi-Channel Distribution

B2B purchasers consume content on different channels, which is the reason why you need to have a multi-channel distribution plan in place. However, the challenge is to deliver content that reflects buyer behaviour and reinforces consistency and personalisation with the experience regardless of all the touchpoints.

According to an Oren Greenberg report, B2B buyers get through 3-7 pieces of content before talking to a sales rep https://www.orengreenberg.com/blog-post/75-b2b-marketing-statistics-for-2024

Some successful delivery strategies are:

- Owned channels, i.e. blogs/website, email marketing, etc.
- 3rd party media such as guest posts, influencer outreach, and media features.
- Paid media (sponsored content, PPC advertising, **LinkedIn** ads, etc.).
- Social media and communities where businesses share insights, interact with industry professionals and build brand awareness.

With AI, companies are able to automate content distribution based on engagement; this means the right message reaches the right audience at the right time.

The Place of AI in the Content Delivery and Targeting

AI is revolutionising how companies disseminate their content by understanding what an audience likes, analysing engagement information, and optimising performance across various channels. Predictive analytics helps marketers automate the placement of content and target content more effectively.

AI-enabled content distribution methods include:

- Dynamic email marketing that adjusts content recommendations based on real-time user engagement.
- An intelligent social media scheduler that posts when the audience is most active.
- Personalised articles, reports or videos recommended by AI

By using AI for management through automation, content is not only automatically sorted and sent out in an effective fashion but can also be constantly re-tweaked and improved, continually ensuring that content performs optimally.

Paid Content Distribution and Retargeting

Paid promotion is a significant enabler for driving increased content reach and discoverability to high-value B2B audiences. With the help of AI, businesses can run focused campaigns that are able to boost engagement and, at the same time, lower costs.

Google is big in the AI marketing game. Their AI-based advertising product lets businesses target users based on signals of intent, making ads more relevant and lowering the cost per click. See the blog by Workshop Digital: https://www.workshopdigital.com/blog/ai-optimized-ppc-google-ads/

Some great paid promotion strategies are:

- **LinkedIn** Sponsored Content for targeting B2B decision-makers.
- **Google** Ads and retargeting campaigns will be used to maintain visibility

during the research phase.
- ABM (Account-Based Marketing) targeting qualifies the recipients, delivering custom content to high-value accounts.

By combining paid promotion with AI-powered insights, businesses can refine targeting, increase engagement, and improve return on ad spend.

Influencer Marketing and Industry Alliances

B2B influencer marketing is an emergent and essential part of an effective content distribution strategy. Companies that work with industry experts, analysts, and thought leaders boost their credibility and reach for content.

According to a report by Ogilvy, 75% of today's B2B businesses are now found to embrace the use of B2B influencer marketing. https://www.ogilvy.com/sites/g/files/dhpsjz106/files/pdfdocuments/Ogilvy%27s%20Influencing%20Business%20Report_1.pdf

Through the use of trusted voices in the industry, companies can boost engagement, demonstrate authority, and create value by nurturing long-term relationships with potential customers.

Measuring Content Distribution Performance

Successful content delivery is an ongoing need that needs to be measured and adapted. Businesses can track performance in real-time with AI-powered analytics platforms, keeping distribution efforts efficient and cost-effective.

Content distribution KPIs can be:

- Engagement data such as time on page, shares and comments.
- Conversion tracking to track the effect of content on leads.
- Click-through rates and cost per lead for sponsored promotion.

By tracking these metrics, companies can optimise their content distribution strategy, make the most of their reach, and increase their marketing ROI.

AI-Driven Content Distribution Is The Future Of B2B Marketing

Content in itself is simply not enough to succeed in B2B marketing. AI-powered and data-driven distribution strategies will be accepted practices, especially for new or established businesses wishing to have their content received by the right audiences wherever they are, whenever they are there.

AI is not going anywhere, and businesses that champion intelligent automation, predictive analytics and personalised content distribution will experience higher engagement, stronger brand authority and better conversion and sales figures. And B2B marketers need to reset how they think about promoting content in a quickly changing digital environment.

Content Performance Measurement and Ongoing Optimisation

It doesn't work to author and distribute content if businesses cannot measure its effectiveness, analyse audience engagement and listen to it in order to shape their tactics. What's more, with the availability of AI-based analytics platforms and performance tracking, B2B marketers can track the effectiveness of their content in real-time, leading them to make data-informed decisions on future campaigns.

A well-informed approach helps ensure content marketing is always in line with business goals while doing everything possible to increase engagement, lead generation, and subsequent conversion.

The Content Performance You Need to Monitor

Tracking isn't one-size-fits-all: measuring the performance of pieces of content depends not only on a team's strategy but also on the goals of their business.

Performance Indicators are:

- Engagement metrics such as page views, time spent on content, and bounce rates.
- Lead generation performance is measured by downloads, sign-ups, and gated content conversions.

- Sales impact metrics, including customer acquisition cost and content-driven revenue attribution.
- Social media interaction such as shares, mentions, and follower growth.

Businesses can use these to monitor which type of content, what subject and what distribution channels are garnering the highest ROI.

AI-Driven Content Analytics and Predictive Intelligence

Artificial intelligence-supported analytics allow publishers to gain a better understanding of content performance through insights such as behaviour patterns, customer preferences and conversion stats. Machine learning based models can pinpoint which content assets are most successful at each stage of the buyer journey, so that marketers can make data-backed adjustments.

In a WordPress VIP survey, 80%of organisations believe that content should be informed by data intelligence, but only 43& believe that their content strategy is data-driven. https://wpvip.com/resource/content-marketing-trends-2024/

AI-enhanced analytics offers the following capabilities and then some:

- Predictive content scoring for which articles or videos are most likely to succeed.
- Automated content discovery by matching audience interests with content products.
- Performance monitoring, in real-time, for perpetual campaign optimisation.

This enables businesses to continually refine their content strategy for maximum reach and impact.

A/B Testing & Performance Improvements

A/B testing, or split testing, is another important method for optimising the effectiveness of your content. By testing variants of a blog article, email subject line or landing page, companies can see which one produces more engagement and conversions.

A case study by Ubisoft used A/B testing to increase lead generation by 12% https://static.wingify.com/vwo/uploads/2018/08/ubisoft-ab-testing-lead-generation_2018-08-29-13-39-12.pdf

Effective A/B testing strategies include:

- Testing new headlines, new visuals, and new content structures until you find the sweet spot.
- Optimising subject lines of email campaigns for open rate.
- Testing offers and CTA placements to boost conversions.

Regular A/B testing means the content is always relevant and engaging, improving overall marketing performance.

Applied AI for Feedback Loops and Sentiment Analysis

The sentiment analysis, with the help of AI, allows businesses to evaluate how the audiences are feeling about their content. From social media conversations, customer feedback and review volumes, companies can derive an understanding of brand perception and audience sentiment.

According to Itrex T-Mobile used sentiment analysis and decreased client complaints by 73%. https://itrexgroup.com/blog/ai-sentiment-analysis-in-customer-service-use-cases-benefits-expert-tips/

Businesses can leverage AI-driven sentiment analysis tools in order to:

- Evaluate emotional reactions to content Find out what topics are striking a chord.
- Monitor how brand perception changes over time against story themes.
- Refine messaging to meet viewers' expectations.

By incorporating AI-powered feedback loops, businesses can be sure that their content is still kept up to date with what customers want as well as with what trends emerge in the industry.

Continuous Content Optimisation for Vigorous Growth

B2B content marketing is not an event, it's a journey of continuous, improvement and evolution of your content. Ongoing optimisation means content strategies can adapt to the needs of the industry, preferences of the audience, and new technology developments.

Some of the key optimisation strategies are as follows:

- Refreshing past evergreen content with the latest in perspective and data.
- Re-imagining top performing content to other mediums like videos, infographics, podcasts, etc.
- Adapting keyword strategies to rising search trends.

Keeping an active content profile helps businesses keep audiences engaged and maximise their marketing power in the long term.

Content Optimisation: The Way of the Future

B2B marketers need to transcend legacy content measurement and adopt AI-based analytics, predictive insights, and agile optimisation. AI doesn't just change the way businesses deliver their content; it changes the way they measure the impact of content and optimises and retargets it in real-time.

Businesses that think differently about measuring content performance use AI for greater insights and do things better consistently will succeed in a rapidly changing digital world.

AI-Powered Search Optimised and Content Discovery

Search is still the foundation of B2B content discovery. The SEO industry, however, has been revolutionised by the introduction of machine learning, voice search, and natural language processing. And in a world where big language models determine how content gets surfaced and interpreted, basic keyword stuffing and ineffective optimisation won't cut it.

AI has elevated SEO into a more strategic, contextual, and predictive exercise. Sophisticated marketers now leverage AI tools not just to identify high-traffic terms but to understand search intent, semantic relationships, and topic clusters that shape how buyers consume information.

Tools like **Clearscope**, **MarketMuse**, **Surfer**, and **Semrush**'s Content AI are no longer just about tracking keywords. They analyse competitors' content structures, assess content scores using topical authority, and give advice on how to increase readability, flow, and comprehensiveness. These tools are not only meant to get a better ranking, but they also dictate the creation of truly useful, stick-worthy content.

Importantly, AI can allow marketers to mimic buyer searches in a more conversational form rather than keywords. This mirrors the way decision-makers actually look, which is through complex questions, long-tail queries, or problem-driven phrases. This optimisation to real-world patterns isn't just about visibility but relevance.

Also pertinent is AI's role in content refresh cycles. Content that might have worked in the past might not work now. AI tools now analyse decaying pages, monitor changes in user interest, and suggest exact updates to keep momentum. I was told by one head of marketing that they now view content as 'old' and need to re-imagine it once it becomes six months old.

With voice search and AI generated summaries on the rise, that means that metadata, Q&A formats and structured content is key. But optimising for meaning to machines is equally important today as human readability is.

The takeaway for B2B marketers is obvious: SEO is no longer just discovering where you rank. It's being the best answer when prospects search. For further clarity, read the next chapter, 'SEO in the Age of AI – Visibility Without Clicks'

Summary:

A good content strategy establishes your brand as a leader in the space, earning trust and legitimacy. Regular, quality content keeps prospects in touch and moves leads along.

- Put out material that speaks to customer pain points, challenges and objections.
- Vary the formats of the content you create in order to appeal to varying sectors of your audience.
- Making your brand an authority in your space.

Case Study: Microsoft's Thought Leadership on AI

Microsoft has successfully positioned itself as a thought leader in artificial intelligence by publishing research reports, hosting AI-focused events, and developing educational content on responsible AI use. Microsoft's AI thought leadership content has been cited in industry reports and has influenced enterprise AI adoption trends globally.

https://blogs.microsoft.com/blog/2024/08/13/unlocking-the-future-of-innovation-the-microsoft-ai-tour/

8: SEO in the Age of AI – Visibility Without Clicks

"The brands that adapt now will have a major edge... Because if AI isn't talking about you, there's a good chance your potential buyers won't be either!"

Corporate Ink, B2B Tech PR and Marketing Agency.

AI has transformed how people discover, evaluate and engage with content. It hasn't replaced search, though, it's fundamentally rewired it. In a universe of zero-click searches, LLM-driven content curation and dark social sharing, visibility is no longer a function of rank alone. The time has arrived to redefine SEO as not search engine optimisation but as 'semantic experience optimisation'.

The New Discovery Landscape

Classic SEO was built for an internet of pages. AI is built for an internet of context. **Google**'s Search Generative Experience (SGE), **ChatGPT**, **Perplexity**, **Claude**, **DeepSeek** and other AI interfaces no longer return lists, they 'compose' answers. And they don't pull those answers from metadata only but from the broader semantic fabric of the web.

That means marketers must shift from optimising 'for search engines' to optimising 'for AI-driven understanding'. The traditional keyword targeting and backlink playbook is no longer enough.

In AI-first environments, users may not even reach your site. Content is being extracted, summarised, repurposed and surfaced in increasingly autonomous ways. You're fighting not just for attention but for meaning.

This shift is not theoretical. B2B discovery now happens outside of direct search and within AI summaries, social shares, and aggregated feeds. The visibility game has moved upstream. https://www.bain.com/insights/losing-control-how-zero-click-search-affects-b2b-marketers-snap-chart/

Search has always been about influence, but now the rules of influence have changed. It is no longer about owning a top-ten ranking. It's about being a reference point in the broader digital narrative. That demands content that is not only technically sound but semantically rich and consistently cited across contexts. https://www.campaignasia.com/article/rethinking-seo-how-googles-ai-overviews-are-changing-the-game/502158

From Clicks to Context

What happens when your prospects never visit your website?

This is the existential question B2B marketers must now confront. In a zero-click world, traditional performance metrics, impressions, CTRs, and bounce rates start to feel increasingly hollow. Success is not just about bringing people to your content but ensuring your content travels to them intact, accurate and imbued with meaning.

Your content must be quotable. Not just shareable in a social sense but functionally quotable by machines. It must be understandable out of context because often, it will be consumed outside its intended environment. Most importantly, it must carry your brand and point of view even when detached from its original source.

We are entering an era in which visibility is determined not only by how users behave but by how machines interpret authority. Language models don't think like us. They infer credibility from repetition, coherence and contextual confidence. If your ideas can't survive de-contextualisation, they are unlikely to be included in summarised outputs.

According to **Intuit Mailchimp**, understanding and optimising for zero-click searches is crucial for maintaining strong SEO performance https://mailchimp.com/resources/zero-click-content/

This places new emphasis on how content is constructed. Semantic clarity, strong subheadings, clear citations, and consistent voice all help AI and human readers extract value. As marketers, we are no longer just creators, we are curators of how meaning is transmitted algorithmically.

The Role of Dark Social

Much of your B2B content never gets attributed. It's shared in **WhatsApp** groups, **Slack** communities, **LinkedIn** DMs and industry Discords. These aren't failures of measurement, they are indicators of trust.

The idea of shared identity is becoming more influential in how B2B buyers assess trust and relevance. Robert Cialdini, author of Influence, later introduced a seventh principle of persuasion: Unity. This refers to the power of shared group membership — tribal affiliation, peer networks, or professional kinship — in shaping judgment. In B2B, Unity manifests in everything from buyer communities to **Slack** groups to invite-only **LinkedIn** forums. As marketers, recognising and reinforcing this sense of belonging can tip the scale when credibility is on the line. https://www.neurosciencemarketing.com/blog/articles/cialdini-7th-unity.htm

Dark social is the digital equivalent of word-of-mouth. It is intimate, peer-validated and profoundly influential. Yet because it evades most analytics tools, it is often undervalued in strategic planning.

Rather than fight it, B2B marketers must embrace it. This means creating content designed to travel, as well as assets that are modular, portable, and valuable in fragments. It means prioritising voice and credibility over format and framing. It means measuring success not just by what gets read but by what gets repeated.

This shift from centralised content to decentralised conversation is one of the most under appreciated evolutions in B2B marketing. It calls for a different type of storytelling, one that anticipates deconstruction and welcomes remixes. When you're thinking appears in a screenshot, a deck, or a **Slack** thread, it's no longer yours alone. That's not a loss of control, it's a sign of resonance.

The best performing B2B content is often not what ranks, but what circulates. In a sense, we are all building folklore for our markets, ideas that move faster than attribution can follow.

Structuring for AI Discovery

To make your content findable and usable by LLMs, think like a knowledge architect.

Large language models parse structure. They look for coherence, consistency, and clarity. If your content lacks semantic markers, it becomes harder for AI to understand and re-purpose. If it lacks credibility signals, it may be skipped altogether.

This is where structured metadata, schema.org standards, and clean internal architecture matter. However, it also extends to how information is layered: FAQs, frameworks, definitions, and 'how-to's' have become essential tools for AI readability.

Think of your content as something that will be read aloud in snippets, paraphrased into summaries, or used to train machine learning systems. It must survive that transformation with your ideas intact.

This is not SEO as we once knew it. It is search, storytelling and software architecture converging. It demands new partnerships within marketing teams - between content strategists, technologists, and data analysts. It also demands humility. We no longer control how or where our content is surfaced. But we can shape how well it is understood.

Brand: The New Backlink

Traditional SEO relied heavily on link-building strategies, but AI systems care more about entity recognition and brand salience than hyperlinks.

The more frequently and consistently your brand is mentioned in relevant contexts, the more AI models associate you with credibility, authority, and insight. A brand becomes a filter for trust.

This elevates the importance of thought leadership, expert commentary, and public-facing clarity. Your brand must exist beyond your website. It must become a recognised semantic object in the AI's conceptual universe.

Every white paper, interview, quote and event you contribute to adds to your cumulative digital footprint. This, in turn, informs the AI's internal model of your relevance. To an algorithm trained on billions of datapoints, authority is not what you say, it's what gets said about you.

To win the future of search, become the answer source, not just a destination.

Strategy for AI-Driven Visibility

To thrive in this new paradigm, B2B marketers need a strategy that acknowledges both visibility and interpretability. But these aren't just boxes to tick, they are interdependent forces that must be planned in concert. Visibility without interpretability risks being superficial. Interpretability without visibility risks being irrelevant. True SEO in the AI era is the practice of making ideas both findable and valuable.

Begin with a content audit, not just for performance metrics but for semantic integrity. Ask: which of your articles, guides or pages are extractable by machines? Which messages make sense when lifted from their source? Can a reader or an LLM understand the point, the brand and the credibility without reading the full page?

Next, map your distribution environment. Visibility now occurs across a fractured landscape in AI summaries, chatbots, content aggregators, curated newsletters and dark social. Where is your thinking appearing? Who is quoting you? Are your insights arriving intact or being diluted in translation? These signals matter more than page views.

Then, build the infrastructure. Layer in metadata, schema.org markup, canonical tags and alt-text. This is not about gaming the algorithm, it's about creating a scaffold of semantic clarity so machines can accurately index, understand and resurface your content. Consider it the architectural plumbing of modern SEO, mostly invisible but foundational.

From there, build authority deliberately, not just with big moments, but with repeated consistent signals. Authority is trained. It is accumulated through citations, re-shares, guest articles, speaking slots, and smart partnerships. Each moment of public contribution is a node in your network of relevance.

Now test it. Ask **ChatGPT**, **Perplexity**, **Claude** or **DeepSeek** questions that your buyers are asking. What comes back? Does your voice appear? Does your thinking echo? This is where your strategy comes full circle, from content creation to AI interpretation. It's a feedback loop that modern marketers must build into their workflow.

Ultimately, this is no longer just about search engines. It's about surfacing in systems of record, in AI co-pilots, in generative feeds and decision-support layers. It's about influencing how markets learn, remember and trust. It is going to be a difficult change, but it's something we will all have to do.

The goal is not just to rank and optimise, it is to be recognised and understood.

Summary:

- AI is reshaping search behaviour: Traditional SEO tactics are evolving as AI-driven platforms like Google's Search Generative Experience and ChatGPT provide direct answers, reducing the emphasis on click-through rates.
- Content must be machine-readable and human-relevant: Ensuring that content is structured for AI comprehension while remaining valuable to human readers is crucial.
- Brand mentions outweigh backlinks: In the AI era, consistent and authoritative brand mentions across various platforms enhance visibility more than traditional link-building strategies.

Case Study: Adobe's AI-Optimized SEO Strategy for B2B Success

Adobe – a major B2B software provider – recently revamped its SEO approach to address the rise of AI-driven search results and zero-click behaviours. The company found that its organic traffic for non-branded searches had plateaued, partly because key product content was too broad.

Adobe's marketing and SEO team made a strategic pivot to semantic content optimization and zero-click visibility. They reorganized content into focused topic hubs and leveraged AI-era SEO tactics to capture attention directly on the search page. Key tactics included: Topical Content Hubs, Featured Snippet Optimization (Zero-Click), and Knowledge Panel & Rich Results.

The impact of these AI-optimized SEO strategies was substantial. Adobe's non-branded organic traffic surged, nearly doubling within 12 months after years of stagnation. https://shubhamdavey.com/adobe-seo-case-study-content-marketing-strategy/

9: Brand Positioning and Messaging Consistency

"A brand is a promise. A good brand is a promise kept."

David Aaker, Brand Strategist.

A B2B brand is not just a name, logo or tagline; it is the company's promise of value that comes with every interaction. That strong brand across touch-points convinces the decision-makers that the organisation is dependable, capable, and even performing and thinking like them. It's a strategic point of distinction in markets where products/ services can seem homogeneous, and it's a reliable foundation for guiding businesses to make complicated purchase decisions.

This chapter dives deep into the aspects of how to build and sustain a consistent brand in B2B, everything from how to create a compelling, Unique Value Proposition (UVP) to how to align internally and use AI to remain relevant in a fast-moving market.

In The Age of AI: The Fundamental Role Of Brand Consistency

Brand Consistency: Why It Matters More Than Ever

Brand consistency isn't just important, it's what sticks in your head when your competitors are all saying the same thing. It influences how organisations build trust, distinguish themselves, and maintain customers over the long term. However, with the rise of artificial intelligence and the way we manage relationships with customers, brand consistency today is about much more than just design, language, or tone. It's more about creating a single experience in every interaction [with the customer] driven by real-time data and automation.

In the modern marketplace, B2B buyers interact with brands via a variety of digital and physical touchpoints - from company websites and social media to AI chatbots and hyper-personalised email marketing initiatives. There's no value if marketing messaging, voice, or values don't agree across channels, causing confusion and dilution of trust. 74% of B2B buyers today cite brand trust as a key factor in purchasing decisions, according to the B2B Thought Leadership Impact Report (2024) by **LinkedIn**.
https://www.edelman.com/sites/g/files/aatuss191/files/2024-02/ 2024%20Edelman-LinkedIn%20B2B%20Thought%20Leadership%20Impact%20Report%20Final.pdf

Companies that combine winning strategies, including consistent branding, can increase their market share by more than 10% annually, according to **McKinsey & Company**'s B2B Pulse Survey. https://www.mckinsey.com/~/media/mckinsey/business%20functions/marketing%20and%20sales/our%20insights/the%20multiplier%20effect%20how%20b2b%20winners%20grow/the-multiplier-effect-how-b2b-winners-grow.pdf

To make this transition, businesses need to adopt AI-powered solutions to keep messaging consistent and adapt in the moment to shifts in customer sentiment, competitive marketplace dynamics, and audience interaction habits.

How AI is Transforming Brand Consistency in B2B Marketing

Traditionally, keeping a consistent brand was achieved through manual processes, such as setting up style guides, brand governance teams and doing periodic audits of marketing material. These approaches still have a place but no longer meet the requirements of the customer in a world where customers want immediate, tailored, and contextually relevant experiences. Brand management is being redefined by AI, automating various processes, offering rich real-time insights and accommodating seamless multi-platform brand harmonisation at scale.

Significant areas in which AI is reshaping brand consistency include:
AI generated and optimised content. AI-powered tools such as **Acrolinx** and **Writer** review content for brand consistency, tone and message

accuracy prior to publishing, guaranteeing consistency through all channels of communication.

Sentiment analysis and real-time feedback loops. AI tools such as **Google** Cloud Natural Language API scan customer interactions across emails, social media, and support tickets, providing immediate insights for optimising and maintaining consistency with messaging and brand voice.

Automated brand audits. AI-based brand compliance platforms, such as **Brandwatch**, run real-time audits on marketing materials, picking up any inconsistencies and keeping global teams in line with brand rules.

AI-based customer engagement support. AI-enabled bots and virtual assistants, like those available from **Salesloft** (Drift) and **Intercom**, are one way to make sure that automated customer interactions maintain a consistent brand voice, enhancing engagement and minimising the likelihood of human error.

By incorporating AI into branding efforts, companies can deliver consistent, data-driven messages at scale and reduce dissonance while also bolstering consumer trust and engagement.

Case Study: How **Salesforce** Uses AI to Strengthen Brand Messaging
One of the most high-profile brands in the B2B technology sector, Salesforce, has used AI to strengthen brand messaging at every customer interaction. With thousands of conversations taking place every day on various channels, such as the website, email marketing, chatbots and social media, Salesforce wanted to create a unified brand experience that appealed to different audience segments.

In order to do so, Salesforce embedded Einstein AI – its in-house AI engine – into its brand communication. Einstein AI reads and understands the customers' interactions, recognises patterns for engagement and personalises content for various personas while staying true to the company's core brand voice. https://www.salesforce.com/artificial-intelligence/

This approach ensures that:

- The brand message remains consistent across the board - from the sales representatives to the marketing teams to customer support staff.
- AI centric content recommendation engines ensure collateral is contextual while staying true to **Salesforce**'s core value offering of a customer led innovation.
- Automated audits scan marketing materials for consistency, ensuring that all assets align with the brand's tone, values, and messaging guidelines.

Leveraging AI as part of its branding strategy has helped **Salesforce** dominate the market thanks to consistent branding, which helps build trust, engagement, and customer loyalty. Businesses leveraging AI-empowered brand consistency tools have witnessed an increase in customer confidence and brand recall, enhancing overall long-term loyalty.

Business Implications of AI-Powered Brand Consistency

A powerful brand is not only aesthetic but also directly affects business growth. The ones that are most successful in doing that are seeing growth of as much as 23% annually, according to **McKinsey**'s B2B Pulse Survey. https://www.mckinsey.com/~/media/mckinsey/business%20functions/marketing%20and%20sales/our%20insights/the%20multiplier%20effect%20how%20b2b%20winners%20grow/the-multiplier-effect-how-b2b-winners-grow.pdf

AI-driven brand management allows businesses to achieve this level of consistency while scaling their marketing efforts globally.

B2B marketers need to think differently about how they approach brand consistency. Old ways won't cut it in an era when AI is driving real-time brand monitoring, automated message alignment, and hyper-personalised customer engagement. Companies that don't leverage AI risk getting knocked out by those that are arming themselves with intelligent branding systems to flow frictionless, data-driven messaging across touchpoints.

The winners will be companies that use AI to monitor and maintain brand integrity, consumer trust and market agility.

Creating a Differentiated Value Proposition in an AI World

The role of the Unique Value Proposition in B2B marketing

A unique value proposition, or UVP, is a key element of a brand's identity that differentiates one business from another. In B2B marketing, involving complex decision making and involving multiple interested parties, having a strong UVP is instrumental in breaking through the noise long enough to develop trust and finally gain conversions. However, as AI continues to redefine what good customer experience means, along with marketing strategies, businesses need to reconsider how they both define and test, as well as communicate their value proposition.

Traditionally, UVP is based on internal ideas, market research, and customer feedback. While these are all still important functions, AI enables businesses to process significant volumes of data, find patterns in customer behaviour, and personalise the message in ways that were previously out of reach of most organisations. Organisations that don't tap into AI-generated insights are in danger of coming up with a UVP that is either too vague, out of sync with market demand, or quickly outdated.

AI Is Changing The Way A Unique Value Proposition Is Developed

The use of AI can, in fact, help businesses measure and optimise their UVP with real-time and predictive analytics. Rather than depending on guesswork or infrequent market analysis, AI-powered technology uncovers what customers struggle with, find elsewhere, or need now.

A few of the main ways AI improves UVP creation are through:

Customer insights with data at the centre. AI-based platforms, including **HubSpot** and **Salesforce**, track customer engagement across digital channels and surface common struggles, trends and purchasing behaviour. Enter AI-powered sentiment analysis, through which businesses can identify what actually is getting through to the audience.

Competitor analysis at scale. AI solutions like **Crayon** and **SEMrush** can sit in judgment on competitor messaging, pricing and customer sentiment, providing immediate feedback for UVP differentiation.

Predictive analytics for market movements. It can process industry reports, stories and economic trends to predict changes in customer priorities.

A/B testing based on AI. There are platforms like **Optimisely**, wherein companies create and test the various iterations of UVP against different types of customers and then utilise AI to measure which of these stories create the most engagement or conversions.

Using AI in these spaces, companies can develop a UVP that is not just unique but also flexible to changing market conditions.

The 'Must have' elements of a Compelling UVP in the AI Era

A good UVP needs to succinctly articulate the brand's promise, what distinguishes the brand from competitors, and address specific customer pain points. While the core principles remain unchanged, AI is influencing how companies craft and refine their UVPs.

A strong, Unique Value Proposition should include:

Clarity and simplicity: The message should be simple and devoid of jargon. Artificial intelligence based tools for sentiment analysis can help keep the language used clear, interesting, and tailored to the audience.

Customer-centric focus. AI helps businesses customise their UVP for individual customer personas using up-to-the-minute insights. Rather than apply a one-size-fits-all approach, AI is able to create value propositions which are dynamic and will appeal to various industries or decision-makers.

Tangible outcomes. A UVP should highlight measurable benefits such as cost savings, efficiency gains, or competitive advantages. AI tools can assess past customer interactions and case studies to identify the most impactful benefits to showcase.

To dig further into emotional levers in B2B marketing, revisit Chapter 2. This section deals with the role played by consistency in reasoning as well as in confidence building.

Real-Life Example: How Adobe Uses AI to Enhance Its Value Proposition
Adobe, the name behind creative (and enterprise) software, has tapped into AI in its branding efforts to ensure its Unique Value Proposition (UVP) stays strong and relevant. **Adobe Sensei**, the company's AI platform, allows customers to scale and optimise digital experiences.

Adobe's UVP is how it enables companies to create great digital experiences. And by using AI, Adobe can customise this message for various industries. For example:

- For creative professionals, meanwhile, Adobe presents its AI-driven tools as a means of turbocharging creativity and working at pace.
- At an enterprise level, the emphasis is changing to AI-powered customer analytics and customisation capabilities.

With AI-powered UVP insights, Adobe keeps updating its UVP and making sure its strategy stays relevant in an evolving market. https://business.adobe.com/uk/ai/adobe-genai.html

How do you create data-driven data for your UVP?

To create an AI-powered UVP, businesses should follow these key steps:

Run AI-powered market research. Leverage AI-based solutions to analyse customer data, monitor competitive placement, and spot trending industry topics.

Spot pivotal pain points and gains. Turn to AI-powered analytics to figure out what the most concerning problems are for the customers and where the brand comes into play.

Use AI to test and perfect messaging. Carry out A/B testing through digital channels to test which of the UVP messages leads to the highest engagement.

Customise UVP for various target readers. Segmentation based on AI offers an opportunity for organisations to customise messaging based on industry type, job role, or geographical market.

Always improve based on real-time data. As market forces shift and evolve, AI tooling provides UVP with real-time feedback to keep it primed and ready.

With AI at every level, companies can keep their UVP data-centric, customer -centric and elastic.

The Future of UVPs in the AI-Powered B2B World
As AI develops, the UVP will evolve as well. Rather than a permanently fixed message, it will become a living asset that changes on the fly based on market changes, customer feedback and competitive movements. Companies that implement AI-supported UVP development will be able to differentiate themselves, form deeper customer bonds, and grow profitably.

Unifying Teams with AI to Preserve Messaging Consistency

Importance of Alignment in B2B Branding

A business' brand is not its marketing materials, logo or ads. It is also determined by how well its internal teams consistently communicate with customers, prospects, and partners. Each touchpoint, whether a sales call or a support chat, adds to an impression about that brand.

When marketing is promising one thing, sales are delivering another, and customer support is telling a third story, then you're taking some trust away from the consumer. This inconsistency has resulted in extended sales cycles, low engagement, and lost revenue.

The CMO reported that businesses with strong marketing sales alignment close 67% more deals while experiencing a 36% increase in client retention. https://thecmo.com/demand-generation/abm-statistics/

Mismatched messages pose problems for companies in a number of different ways. If, in our marketing, we're promoting a solution one way, but our sales team is talking about it in a different way, customers will get confused. Where there is a misalignment between sales and marketing, lead-to-sale conversion rates may decrease. If your customer support people aren't up-to-date on the latest messaging, they could provide some old or even conflicting messaging and frustrate prospects. Skewed communication among various teams may affect the company's reputation and can reflect an unorganised look for the company.

Internal Brand Training and Knowledge Management powered by AI

AI-powered solutions are helping B2B companies ensure brand alignment across internal teams by automating brand training, standardising content, and providing real-time guidance. AI can audit, analyse, and refine messaging across various channels, ensuring every touchpoint reinforces the same brand story.

Retraining, re-communication, and re-affirmation of what they've learned need to be a continuous cycle. Static training manuals and occasional brand workshops are giving way to real-time, interactive training that's tailored to employees' needs on AI-based learning platforms.

AI-infused learning platforms like **Spekit** provide micro-learning for employees to get brand positioning and tone that is engaging and easy enough to remember. AI chatbots are used as internal knowledge assistants: employees can instantly pull the latest brand messaging guidelines, ensuring consistent communication between teams. AI-powered coaching tools listen to call recordings between sales and support staff and provide feedback on their tone, message alignment and customer sentiment, enabling them to develop their communication skills.

Sales and Marketing Alignment with AI Benefits

Sales and Marketing 'misalignment' is a known factor that B2B organisations often simply don't talk about. Marketing teams develop messages and content to win over leads, but if sales teams fail to use that content, customers could end up with out-of-date or contradicting info. AI is closing that loop by having real-time messaging recommendations placed into sales workflows.

AI-powered conversation intelligence tools like **Gong.io** and **Chorus.ai** listen in on sales calls and coach in real-time to keep reps on track. AI-powered content management platforms like **Seismic** and **Showpad** help sales teams surface the most current, marketing-approved materials so they don't use outdated messaging. AI insight suggests which messaging connects most with individual targeted customer segments so teams can iterate on messaging and personalisation strategies.

Consistency of Customer Support in an AI world

AI is also helping customer service teams to support brand consistency. Support agents need to deliver correct, consistent and current responses, especially in industries where technical knowledge is a critical part of a customer conversation.

AI customer service tools like **Intercom** and **Zendesk** AI work behind the scenes for real-time messaging suggestions, so support agents use the most appropriate language and tone in their responses to user enquiries. Sentiment analysis powered by AI enables support teams to make and adapt responses according to customer feelings, allowing them to present the same supportive brand tone in a more compassionate light. AI-generated FAQs and knowledge bases provide consistently updated self-service support material that is easily accessed by customers and staff.

The Future of AI for Message Consistency and Internal Alignment

AI is changing, and we are only going to see more of it in terms of maintaining a cohesive messaging strategy across teams. Other AI-generated innovations on the horizon include tools that analyse employees' spoken words to ensure that they're using the right tone when interacting with customers and even suggesting changes in real-time during such interactions. AI-powered generative assistants for authoring sales scripts and customer emails that guide authors to write in a brand-compliant way. Multi-language AI translation tools that preserve message consistency in global markets take cultural differences into account.

AI is the Connective Tissue for Brands Internally

Keeping messaging unified and consistent on the marketing, sales and customer support side is becoming increasingly difficult, but AI is helping keep that message cohesive and scalable. Businesses that invest in AI-driven brand governance, real-time coaching, and automated knowledge management will win big as the B2B world becomes more and more digital.

Through AI, companies can alleviate inconsistencies across messaging, make communication more efficient and ensure that every customer experience supports a unified brand experience.

Evolving Brand Position for a Shifting Market

Imperative for Ongoing Brand Evolution in B2B

Brand positioning necessarily shifts as markets, customer needs, competitive pressures and technology move on. Brands that do not adapt their positioning will become irrelevant and will be eaten up by the competition. While brand continuity is key, companies must also remain adaptable in order to stay relevant to the audience they're trying to reach.

If you want to get more into the psychology of what makes a B2B buyer tick, then check back in Chapter 2. Here, the discussion is about how consistency of message can help both rational testing and emotional confidence.

According to research from **Gartner**, most CMOs are expecting a positive impact from AI. https://www.gartner.com/en/marketing/research/digital-iq-index-genius-brands As AI is changing entire industries, businesses that incorporate AI into their brand positioning strategies will stand a better chance of retaining a unique market space and tailoring their messages accordingly to what the market demands.

There are several reasons why your brand positioning will need to be altered on an ongoing basis. Buyer preferences change as industries mature and new technologies become available. Competition also heats up as new competitors enter the market, competing with established players and destroying differentiation. Going global also involves customising communication for various cultural, economic and business settings. Digital transformation is changing the way businesses communicate with their customers, which means businesses need to align their positioning with changing communication channels and engagement strategies.

Artificial intelligence is becoming an essential tool in helping businesses refine their positioning dynamically. AI-powered solutions provide real-time insights into market trends, competitive movements, and customer sentiment, allowing companies to adjust their messaging in a data-driven manner.

How AI Helps Adapt Brand Positioning

Artificial intelligence redefines brand positioning by providing predictive analysis, competitive intelligence, and real-time market intelligence. Businesses that weave AI into their strategic planning can expect to be able to predict market moves, listen to customers, and keep their messages unique from those of their competitors.

AI based Market and Trend Analysis

Understanding market trends is a key factor in successful brand positioning. AI tools can process gigantic volumes of industry data in real-time, spotting trends and predicting changes before they become obvious to everyone. Instead of the old static market research reports, companies can now refine their strategy in real time based on what people are doing right now.

AI trend-tracking platforms like **Brandwatch** and **Crimson Hexagon** monitor online conversations, media reports, and industry discussions to detect early indicators of market changes. These are the kinds of tools that help businesses shift their messaging proactively and remain ahead of the industry curve.

Search analytics tools that employ AI, like **Google** Trends and **AnswerThePublic**, monitor fluctuations in buyer search intent as to which topics are increasing in relevance, enabling brands to sharpen their messaging according to topics that their key audiences are developing a deeper interest in.

Competitive Positioning Enhancement by AI

Competitive differentiation is essential for effective brand positioning. AI-powered competitive intelligence platforms provide businesses with a real-time window on their competitor's messages, pricing strategy and market positioning. These are the kinds of insights that can help brands further position themselves and build value propositions that stand out.

AI-driven competitive intelligence solutions like **Crayon** and **SEMrush** track competitors' content, product launches, and pricing strategies, which allows businesses to know if they're in line with the competition or where they can differentiate themselves.

Sprinklr and **Meltwater** are both AI-based social listening platforms that analyse customer sentiment surrounding competitors while also providing businesses with the ability to spot market gaps and areas to improve.

Customer Sentiment Analysis and AI-Generated Messaging Adjustments

A company's brand positioning has to correspond to how its customers think about them. AI-driven sentiment analysis tools help companies monitor customer sentiment, track brand perception, and fine-tune messaging to meet what their audience expects. These tools analyse customer reviews, social media conversations and comments on online forums to give real-time insight into how a brand's positioning is catching on with a variety of groups.

AI-powered sentiment analysis tools such as **Lexalytics** assess customer emotions and see if your company's messages are genuinely making an impact.

Natural language processing tools evaluate customer conversations to determine if brand messaging resonates with audience attitudes. These insights are what enable businesses to optimise their messaging in real-time to keep positioning as relevant to the marketplace as possible.

Content Personalisation and Brand Adaptation using AI

AI also allows brands to customise their messaging without losing brand positioning. Personalisation engines that leverage AI decode customer data to ensure that communication is aligned with the overall brand strategy.

Content personalisation platforms like **Persado** and **Adobe Sensei** leverage machine learning to test variations of brand messages across multiple audiences segments, to find the best-performing messaging. Such platforms enable companies to evolve their positioning continually, keeping their messaging fresh to resonate across various industries and types of customers while maintaining consistency.

Case Study: How HubSpot Uses AI to Evolve Brand Positioning Over Time
HubSpot, a company that sells inbound marketing and sales software, is a case in point for how a company can tap AI to further polish its brand positioning as customer priorities and market dynamics change. Leveraging AI-powered insights, HubSpot has pivoted its messaging while preserving a clear and familiar brand voice.

AI-enabled sentiment analysis tools are used by HubSpot to follow trends in the industry through customer conversations, competitor positioning and content performance. This gives the company the ability to fine-tune its messaging based on the most innovative marketing automation and CRM strategies. For example, AI-powered customer engagement analytics enables HubSpot to determine what types of content and messaging are most appealing to sales-qualified leads and communicate more personally in a manner that reflects changing customer preferences.

By incorporating AI into its branding efforts, HubSpot has kept its competitive edge in the rapidly evolving marketing technology industry so that the positioning of the brand is relevant to today's business needs. https://blog.whitehat-seo.co.uk/hubspots-ai-2025-report

The Future of AI in Brand Positioning

With the rise of artificial intelligence, its importance regarding brand positioning is likely to become even more apparent. This will further unlock the potential for companies to predict industry trends more accurately than ever before, generating data-led, on-the-fly recommendations on how to change brand messaging. AI driven sentiment tracking will provide immediate feedback on how customers are reacting to brand positioning, enabling companies to iterate quickly. Brand audits will be automated to maintain the consistency of messaging and enable market-driven shift in order for brands to remain competitive.

Companies that invest in AI-powered brand positioning strategies will have a huge competitive edge by outpacing market changes and continuing to build strong emotional ties with their audiences. Brands that can strike the right balance with consistent branding practices and are adaptable will most likely succeed over the long term within B2B.

Using AI to Ensure Message Consistency and Brand Alignment

AI, Ensuring You Keep on Message

It's a tough task for many B2B businesses to keep their brand messages the same throughout all of the 'touchpoints'. The number of different online and offline channels used by businesses to talk and listen to customers via email marketing, social media, website content, sales calls, and customer support adds to the complexity. Misaligning brand messages between these channels can cause confusion, lack of credibility, and undermine the overall punch of the brand.

Brand harmony necessitates that marketing, sales and customer service work together, so that each echelon articulates a uniform value proposition in a uniform manner. Without structured alignment, businesses risk presenting fragmented messages that fail to build trust with their audience.

Artificial intelligence has removed the guesswork for businesses looking to standardise and improve their messaging across platforms. Artificial intelligence powered tools keep brand messages intact out in the field, adjusting them to different engagements on the fly.

AI-Enabled Content Moderation and Governance

Brand messaging is increasingly tracked by AI in audits that keep the brand message on voice. AI-driven platforms can scan marketing materials, emails, sales presentations, and customer interactions to ensure they align with the company's approved messaging framework.

AI-powered content moderation tools such as **Acrolinx** and **Grammarly** Business help organisations maintain brand consistency by automatically reviewing content for tone, message alignment, and compliance with brand guidelines.

AI-driven governance systems also help prevent off-brand messaging from being shared with customers, ensuring a high level of consistency across all content.

AI's Role in Messaging Coherence and Brand Alignment going forward

Brand management with AI is likely to increase in significance as tech progresses. Businesses that adopt AI brand alignment solutions can maintain brand control over large scale and numerous teams, geographies, and touchpoints, providing business efficiency and scalability.

AI-driven predictive analytics will allow companies to anticipate future brand inconsistencies before they ever occur so marketing, sales and customer service teams can keep an eye on potential misalignments. AI will do sentiment analysis in real-time, feeding back to companies on what's going on with customers and in the market in an evolving fashion so they can constantly adjust the message. Multilingual messaging tools with AI underpinning will allow organisations to take advantage of a consistent brand position globally while adjusting comms to local languages and cultural nuance.

Building Brand Architecture in a Fragmented World

As B2B companies scale and diversify, managing a portfolio of products, services, or sub-brands becomes a strategic imperative. A transparent brand architecture prevents marketers from getting lost and supports cross-selling and the establishment of a consistent brand experience in multiple markets.

The different types of brand architecture are:

Branded House: All products and services are branded under the one master brand with sub-brands as derivatives of the parent brand.

House of Brands: Each product or service is branded independently

Hybrid: Some sub-brands can maintain their identity while others use the parent brand.

The Role of AI: Implement AI-powered brand sentiment analysis to help guide consolidation or expansion of the brand. You can use AI tools such as **Qualtrics** or **Brandwatch** for a brand portfolio audit and evaluate the consistency of perception, channel clarity, and redundancy.

Summary

A strong brand positioning strategy and consistent messaging are fundamental to building trust, recognition, and long-term customer relationships in B2B marketing. Businesses that align their messaging across all channels and maintain a clear value proposition create stronger engagement and differentiation in competitive markets.

- Ensure that messaging aligns with your brand's core values, mission, and long-term vision to create an authentic and compelling brand identity.
- Maintain consistency across all marketing, sales, and customer service interactions to reinforce brand trust and credibility.
- Leverage artificial intelligence to monitor, refine, and adapt your brand positioning in real-time, ensuring that your messaging remains relevant in a changing marketplace.
- Regularly conduct audits and performance analyses to evaluate how well your brand positioning resonates with customers and make adjustments as needed.

Example in Action: GE's Brand Positioning

General Electric (GE) has established a brand positioning strategy that is deeply rooted in two key pillars: trust and innovation. Since its inception, GE has been synonymous with cutting-edge technological advancements and reliability. This strategic focus has enabled the company to appeal to a wide range of industries, from energy and healthcare to aviation and digital solutions.

GE consistently communicates its brand message across all platforms, reinforcing its role as a forward-thinking, solutions-driven company. By maintaining alignment between its innovation-driven messaging and customer expectations, GE has successfully built a brand that is both aspirational and dependable, ensuring long-term customer loyalty and industry leadership.
https://www.marketingexplainers.com/ges-marketing-strategy-explained/

10: Customer Advocacy and Referral Marketing

"Customer experience is the new marketing."

Jay Baer, Customer Experience Expert.

Customer advocacy has become a key tactic for B2B companies looking to create long-term relationships, build trust, and drive revenue growth. Unlike conventional marketing, advocacy identifies happy customers as brand ambassadors, able to sway peers and influencers in their respective professional circles.

In today's digital world, where data rules, customers advocating on one's behalf have an out-sized impact on B2B purchase decisions. At the end of the day, decision makers are most influenced by their extended network of referrals, testimonials and peer recommendations to validate vendors before engaging with them. Advocacy programmes are evolving yet again, thanks to new age capabilities in Artificial Intelligence that empower businesses to find, target and quantify the impact of their brand advocates with more precision.

Jay Baer, a recognised marketing and customer experience expert, tells us that customer advocacy is the single most potent factor in growth because it means your customers turn into marketers. This reflects a broader trend towards more organic, trust-based B2B marketing.

According to **Forrester**, more than 90 percent of B2B buyers trust industry peers to help guide their company purchase decisions, meaning peer recommendations have some real credence. https://www.forrester.com/blogs/who-do-b2b-buyers-trust/

AI is transforming the way that companies discover and cultivate advocates, making it easier and cheaper to have referral and advocacy programmes that are real and scalable. These days, if you're still working with a spreadsheet and generic, all-purpose emails, you're not scaling; you're stalling. AI is what puts advocacy in motion.

How to Create an Effective Customer Advocacy Programme

An effective customer advocacy program turns happy customers into advocates, who, in turn, advocate for a company's products or services. In B2B, where purchases are more complex with multiple stakeholders, advocacy speeds up the trust-building process and even changes the way people buy. Companies that operationalise their advocacy programs build stronger relationships, have higher retention and can turn advocacy into a defensible competitive advantage.

Customer advocacy programme success hinges on getting your objectives right, offering the right incentives, and having AI pick out and engage the brand advocates who matter at scale.

Identification of Intended Advocates

Your customer advocates are your happiest, long-term clients. They are the ones who derive measurable value from what you offer. They might be repeat customers, an avid recommender, or clients who've been featured in good news stories in a case study or testimonial. With AI-enabled customer relationship management (CRM) tools, a brand can now analyse the behavioural data, sentiment and engagement history to identify the right brand advocates.

With AI-driven analytics, companies can predict which customers are most likely to become advocates and reach out to those customers.

Engaging Advocates and Offering Incentives

Having found the potential advocates businesses ought to attract them with powerful incentives. However, rewards should be matched to the advocates motivators; some will want financial compensation, and others will want professional recognition or networking opportunities.

Some successful engagement strategies are:

- Access exclusive goods, beta tests & innovation forums.
- Industry awards, speaking gigs, guest blog posting.
- Monetary rewards in the form of discounts, free upgrades or outcome-based incentives.

A **Forrester** study shows how **Influitive** found that implementing a structured advocacy programme led to a five percent increase in customer retention over three years. https://influitive.com/blog/forrester-study-influitive-roi/

For long-term retention, advocacy programs need to be personal to the customer, who should feel valued and incentivised in a way that matters to them.

Scaling Advocacy with AI and Automation

Historically, customer advocacy was done through old-fashioned manual outreach and tracking and couldn't be scaled easily. AI and automation now allow companies to successfully manage advocacy programmes by:

- Monitor customer sentiment and engagement in real-time.
- Referral tracking and rewards automation.
- Discovering the best messengers in industry segments.

AI not only makes advocating easier, but it also means businesses can track and fine-tune program effectiveness with real-time data.

Customer advocacy is not just a soft marketing approach, it's a structured, data-driven strategy that makes impact on revenue growth. Personally, I have used customer advocacy as a secure, cost-effective, and high-impact driver. Companies mastering AI to spot, engage and manage advocates will establish a long-term competitive edge and reduce customer acquisition costs.

Businesses that redefine what it means to advocate, expanding beyond referrals and incorporating AI insights, will forge deeper connections, grow faster and establish themselves as leaders in the AI era.

Creating a Referral Programme

For B2B companies, a well-structured referral program is one of the most cost-effective ways to get new customers. Referral marketing takes advantage of the trust that already exists between business professionals. Word-of-mouth is one of the most authentic ways to consider the value of a product advertised, as consumers rely more on brand advocates than the brand itself.

Artificial intelligence now helps referral programs become more effective by helping businesses find the right referrers, manage incentives automatically, and monitor the performance of referral-driven leads in real-time.

The Business Case for B2B Referral Schemes

Referral marketing in B2B settings is highly effective because purchasing decisions are often complex and require validation from multiple stakeholders. Decision-makers are more likely to trust referrals from industry peers than traditional marketing claims.

A study by **McKinsey** revealed that consumer-to-consumer word-of-mouth generates more than twice the sales of paid advertising across various industries. https://www.mckinsey.com/capabilities/growth-marketing-and-sales/our-insights/a-new-way-to-measure-word-of-mouth-marketing

These findings illustrate how businesses must create formal, measurable and AI-powered referral programmes to maximise customer acquisition.

Core elements of a Successful Referral Programme

A good B2B referral scheme should be strategic, automated, and quantifiable. The key ingredients to make a programme generate high quality referrals are:

- Clear incentives – Referral rewards should align with the value customers seek, whether in the form of discounts, exclusive services, or brand recognition.

- Seamless participation – The process should be simple and frictionless, allowing referrers to submit leads with minimal effort.
- Automated tracking – AI-driven platforms can monitor referral performance, ensuring that businesses identify the most effective referrers and reward them accordingly.
- Multi-tiered incentives – Offering tiered rewards, such as higher incentives for repeat referrals, encourages ongoing engagement.

According to Referral Rock, 71 percent of B2B companies experience higher conversion rates from referrals than other customers. https://referralrock.com/blog/b2b-referral-marketing-statistics/

By integrating these factors, companies can create referral programs that enable long-term growth and new customers.

Using AI to Enhance Referral Programme Performance

AI is revolutionising the way in which we do referral marketing by enabling companies to:

- Use historical engagement data to find your best referrers.
- Perfect your referral incentives by determining which incentives generate the most participation.
- Automate outreach and follow-ups so you don't have to follow up with advocates manually to keep them engaged.
- Track and optimise what referrals are driving with AI-powered attribution models.

Through its automation, analysis, and optimisation capabilities, AI allows companies to generate better leads at a lower acquisition cost.

Referral Marketing in the Age of AI

A well-designed referral program is no longer a nice-to-have, it is a critical piece of B2B customer acquisition today. Businesses utilising AI to improve targeting, automated processes, and tiered referral incentives will build a scalable and efficient growth machine.

In the age when trust is the ultimate currency for a purchasing decision, companies that incorporate AI-driven insights into their data-powered referral marketing strategy will experience lasting success.

Measuring Advocacy Success

Impactful customer advocacy and referral marketing must be measured and monitored constantly. Without thorough performance monitoring, companies run the risk of investing in programmes that aren't making an impact. With AI-powered analytics, companies can now measure the immediate value of advocacy efforts in real-time and make sure that referral programs are a part of long-term business strategy.

Effective measurement involves tracking key performance indicators that reflect customer engagement, conversion rates, and return on investment. AI-powered tools enhance this process by providing deeper insights into audience behaviour and identifying areas for improvement.

Key Metrics for Measuring Advocacy and Referral Success

Measuring the right KPIs enables companies to optimise and improve customer advocacy and referral programs better.

Referral Conversion Rate – This is the percentage of leads referred who become customers. A positive value indicates strong advocacy performance.

Customer Lifetime Value (CLV) of Referred Customers - Strives to measure whether referred customers are more sticky and valuable in the long run than customers acquired through other channels.

Net Promoter Score (NPS) Trends – Monitors if customer perception changes as a result of advocacy.

Cost Per Acquisition (CPA) for Referrals – Measures how much it costs to acquire a customer via referrals vs other paid channels.

Engagement Metrics – Measures how frequently customers engage in advocacy activities (e.g. submitting a review, testimonial, case study, etc.)

These metrics allow businesses to quantify the impact of advocacy and make data-driven adjustments to their programmes.

Leveraging AI for Better Referral and Advocacy Tracking

Advocacy and referral measurement are being improved with AI, which is now delivering real-time customer interaction and engagement insight to attribution modelling. Core AI powered features include:

Predictive analytics - AI evaluates historical data to determine what referral sources will drive lifetime high-value customers.

Automated tracking - AI tools that track and analyse referral patterns, helping to prevent tracking failures.

Sentiment analysis - Machine learning programs evaluate customer sentiment by reviews, social media and support interactions as a measure of advocacy effectiveness.

Supporting evidence:
Salesforce research indicates that companies tapping AI to monitor advocacy and referrals achieve a 32% higher rate of customer retention. https://www.salesforce.com/resources/research-reports/state-of-marketing/

Forrester predicts that by 2026, 70 percent of B2B organisations will use AI-powered sentiment analysis to measure impact on customer advocacy. https://www.linkedin.com/pulse/ai-personalization-future-b2b-sales-2025-kensium-wvc2e/nsium-wvc2e/

AI-powered tracking not only improves productivity but also helps businesses to adapt and tailor their advocacy efforts using data-led insights.

Long-Term Success Through Data-Driven Advocacy

It is important to measure the effectiveness of advocacy and referral programmes in order to maximise their impact on selling and growing the business. By using analytics powered by AI, businesses gain greater visibility into customer interaction, so they can more accurately monitor referrals and fine-tune their strategy by checking results.

Companies that implement AI tech to track advocacy will receive a competitive edge by lowering customer acquisition costs, amplifying referral success, and improving customer loyalty. Those who can think differently and adopt data-driven insights will drive the long-term success of their advocacy programs.

With peer-to-peer purchasing now the order of our day, communities have become an impactful engine of trust within B2B marketing. What buyers are after now is not simply vendor validation but practitioner, influencer and user group validation of those same practitioner journeys.

Community-led growth (CLG) is the practice of leveraging customer, partner, and user communities to boost awareness, engagement, user retention and advocacy. It's turning the marketing funnel upside down: rather than push messaging out, it creates room for people to engage, learn, and share on their own.

The most successful B2B brands, and increasingly B2C brands, are building ecosystems, not products. **Salesforce** has Trailblazers. **HubSpot** has its HubFans. **Gainsight** operates its Customer Success Network. These communities are creating content, answering questions from peers, advocating to the public, and helping to lower the support load, all while digging in emotionally to who you are as a brand.

Marketers manufacture this up overnight. But they are able to invest in the conditions that allow it to flourish:

Establish your safe spaces: Slack groups, forums, **LinkedIn** groups, or live events where your peers can talk frankly without feeling like they are being sold to.

Amplify voices: Highlight community champions with events, webinars and social takeovers.

Provide exclusivity: Early access, beta programs, advisory boards, or certification tracks that show the member is special.
Curate for relevancy: Make content and conversations member-led, not only moderated by marketing.

The benefits are profound. Brands enjoy higher retention, lower acquisition costs (through referrals) and stronger organic reach. In our low-trust, high-competition age, a healthy community is not a nice-to-have but a strategic must.

Summary:
Happy customers are your greatest marketers. Referral programs are how many of your clients can become your best brand advocates.

- Create refer-a-friend or customer advocacy programs with incentives.
- Highlight customer success stories to attract new prospects.
- Foster strong relationships to encourage loyalty and advocacy.

Example in action: Slack's Evangelist Program
Slack encouraged its customers to become brand advocates by sharing their success stories. By incentivising referrals and highlighting customer achievements, Slack turned loyal users into powerful marketing assets, driving organic growth through word-of-mouth.
https://medium.com/empathy-for-change/slack-case-study-5833a57289d2

11. Engaging Prospects: Creating Buyer Journeys

"Make your customers the hero of your stories."

Ann Handley, Digital Marketing Pioneer

Selling effectively to B2B prospects is a complex art that needs equal parts of great content, data-driven personalisation, and a big emphasis on relationships. Unlike B2C sales cycles, B2B buying cycles are long and complex, often requiring several decision-makers and bearing with them significant financial and logistical consequences. To navigate such a diverse landscape, marketers need to have a multi-layered strategy in place so they can meaningfully engage prospects and offer them valuable knowledge at every step of their journey.

The world of lead generation has changed, and now, the ability to know who to talk to has been revolutionised through AI, machine learning, and predictive analytics. Such technologies enable marketers to personalise interactions, deliver at the right time, and predict prospect behaviours like never before. Indeed, the transformation to digital and remote engagement is the new normal, as you can see from this article by **McKinsey**:
https://www.mckinsey.com/capabilities/growth-marketing-and-sales/our-insights/these-eight-charts-show-how-covid-19-has-changed-b2b-sales-forever

Future Strategies for Lead Nurturing over the long term

Lead nurturing in B2B marketing is an investment to establish continuous relationships. B2B lead nurturing should help to guide potential customers through difficult purchasing decisions. Nurturing is a delicate balance of good content, personalised communication and timing.

Content as a Core Tool for Engagement

Content is the foundation of any successful lead nurturing strategy. It is an introductory touchpoint, an educational tool for prospects, and, at the same time, directs the prospects' decision making process. In order to make the most impact, B2B marketers must mix up the types of content they're producing and make sure it's tailored to every stage of their buyer's journey.

Expanding Content Types and Formats for Comprehensive Engagement

Thought leadership content and original industry reports help position your brand as an expert in your space. This content type is so useful in giving prospects a sense of what you offer; it helps people make more informed decisions about who to work with – no small feat in solidifying your reputation as a trusted source. Four out of five modern B2B buyers crave thought leadership content, with 64% of them preferring thought leadership to promotional content. https://www.edelman.com/sites/g/files/aatuss191/files/2024-02/ 2024%20Edelman-LinkedIn%20B2B%20Thought%20Leadership%20Impact%20Report%20Final.pdf

Visuals like infographics and even data visualisations are great for breaking down complicated data in an easy-to-understand way. In the B2B space, data-rich topics can be challenging to navigate; thus, making your information visual can help add another level of understanding and engagement.

Interactive content such as assessments, quizzes, and ROI calculators get prospects involved by delivering an instant, personalised brand experience that is relevant to them. This tactile experience can enhance engagement by as much as 2X since quizzes and calculators receive more than double engagement than static content, according to **Embryo**: https://embryo.com/blog/interactive-content-statistics/

Podcasts and webinars are also great tools for prospects to interact with your brand on their own terms. Webinars can either be live or recorded, with a combination of live interaction and viewing on demand. It's perfect for deep-dive topics and gives prospects an immediate interaction.

Providing Value in a Consultative Manner

In B2B, aggressive sales tactics can work against you. A consultative, respectful pace and helpful approach help us build trust and move prospects closer to conversion. In reality, 77% of B2B buyers report that their latest purchase was very complex or difficult because they ran into a buying process that was too complex, including meetings and decent sales pitches https://saleslion.io/sales-statistics/77-of-b2b-buyers-reported-that-their-latest-purchase-was-either-difficult-or-very-complex/

Clients today want a more private experience rather than high-pressure sales tactics.

Buyer Experience (BX) is UX, CX and brand promise aligned in one orchestrated journey. The point is not just how buyers feel but also how each interaction contributes to trust, loyalty and action. It is the new 'brand'.

AI-powered journey orchestration platforms such as **Insider** or **Dynamic Yield** could lead to consistent BX across owned and paid channels.

A soft-sell approach focuses on value and support rather than straightforward selling which makes people a little more than happy exploring your brand without feeling like you're pushing a sales agenda.

B2B buying cycles can be long with multiple decision makers, so accommodating the buyer's timetable is critical. By letting prospects engage on their terms, marketers can build a stress-free experience that encourages real, authentic engagement.

Personalisation is crucial for B2B interaction, as buyers demand tailored communication that fits their industry, role, and stage in the buying cycle. And with AI and data analytics, marketers are able to create hyper-personalised campaigns that speak to each individual prospect.

Techniques for Hyper-Personalisation and Micro-Segmentation

AI and machine learning enable marketers to perform real-time content personalisation, tailoring website content, email, and ads to the prospects in response to their individual behaviour and interests. This is all to make every touchpoint meaningful and timely.

Behavioural triggers do just that: they automate the sending of a message based on a certain behaviour, be it a page visit, download or email interaction. Automated segmentation also helps content remain relevant and topical without relying on continual manual tweaks. https://www.linkedin.com/advice/0/what-benefits-using-marketing-automation-segmentation

Predictive analytics also enables marketers to know in advance what kind of content a prospect might find valuable next. By analysing engagement data trends, marketers can offer value in the right stage of the buyer journey before a prospect even realises it might be necessary. https://6sense.com/platform/predictive-analytics/what-is-predictive-marketing/

Converting leads from awareness into active participation is even more necessary in B2B marketing. It's about crafting a journey that earns trust, adds value over time, and positions your brand as a solution to potential customers' individual challenges.

Every stage of the buyer's journey should be carefully considered to nudge a lead further down the path to ultimately make a purchase, and there are different tactics and content formats that can be tailored at each stage.

Internal and External Data Sources Utilisation

By knowing where B2B buyers are looking for information, marketers are better able to place their messaging in the right place at the right time. The average B2B customer consults internal and external sources, comparing what the vendor presents with independent, third-party perspectives.

Leverage Internal Resources For Easy Access and Value

A repository of resources on your site, case studies, white papers, FAQs, how-to guides, etc., enables prospects to discover your product on their own. This hub can be a one-stop shop for content.

For B2B, prospects may need high-level and highly detailed technical information. Mix it up with technical guides, practical use case studies and product solutions to make sure all personas can find the information they need.

Establishing Credibility Via the Network

Many prospects copy and paste review websites such as **G2**, **Trustpilot**, and **Capterra** into **Google** to see what feedback is being said about a service. Through the promotion of satisfied customers to leave reviews on third-party review websites, brands can establish credibility in these 3rd party review platforms.

Collaborating with industry influencers and getting featured in established publications adds credibility. Especially influencers, who can contact directly with your target niche audience members to give them organic and trustworthy recommendations.

The use of AI and Machine Learning for Prospect Engagement

AI/ML integration in B2B engagement strategies is game-changing, making it possible for companies to personalise the delivery of content and predict what prospects need and when.

AI can analyse behavioural data in real-time, discerning trends, which makes hyper-personalisation possible. They can serve content at exactly the time prospects will engage.

Predictive analytics taps into data, both past and present, to predict the probability of certain actions, which allows marketers to be proactive in their approach.

Summary: Engaging prospects requires personalised, value-driven approaches. Effective engagement involves understanding prospect needs and leveraging the right channels.

- Use data to personalise content and outreach strategies.
- Experiment with new engagement channels, such as interactive content.
- Build trust by offering value before asking for a commitment.

Case Study: Adobe's Personalised Campaigns

Adobe used AI-driven tools to deliver hyper-personalised campaigns. By leveraging customer data, Adobe crafted individualised experiences that spoke directly to the needs and preferences of each prospect. The result was higher engagement rates and stronger conversion metrics. Adobe's AI-driven Sensei platform enabled dynamic content customisation, enhancing engagement and conversion rates. https://business.adobe.com/uk/customer-success-stories/british-telecom-case-study.html

12. Channel Strategy that Scales

"The essence of strategy is choosing what not to do."

Michael Porter, Strategy Expert

In B2B marketing, a robust channel strategy is essential. It's just basic relevance and growth. Contrary to their B2C counterparts, B2B buyers live in high-stakes, elongated sales cycles and require the consensus of many decision-makers. This ambiguity means that maintaining alignment, messaging, and timing is crucially important to ensure the stream of electrons is focusing on the same sort of scenarios. A robust channel strategy doesn't only back up your marketing; it runs it.

Channel strategy at its core is about alignment: the alignment of the message with medium, of content with context, of the brand with the buyer. It's all about making every touch point, from a social post, webinar, white paper, direct mail piece, email, etc., contribute meaningfully to the overall experience.

B2B Channel Strategy Basics

In order to create a scalable channel strategy, marketers must start by rethinking and relinquishing their "plan to distribute content" and instead adopt a "plan to design journeys". The goods are not the end; they are the vehicle. Every asset must be created with the intent to address a need at an explicit stage of the buyer's journey. When you understand how buyers move, not just linearly, but back and forth between stages, it enables marketers to predict needs, conveying that information in ways that guide rather than push.

Here's where strategy comes into play. It starts with understanding your audience, not only their identity but also their behaviour. Segmenting your audience and developing buyer personas is the bedrock. These identities are developed based on role, industry, intent signals, challenges and means for content consumption. This is where AI and machine learning come in, enabling these personas to be dynamic, constantly updating to reflect new customers and how they engage with a product or service.

A **Gartner** survey reveals that 86% of B2B customers expect companies to be well-informed prior to interactions. https://www.gartner.com/en/newsroom/press-releases/2022-05-13-gartner-survey-reveals-b2b-customers-expect-companies-to-be-well-informed

Choosing the Right Social Media Platforms

Knowing your audience is only the start. Then comes the question of where to locate them. This does not imply being geographically everywhere; instead, it is about being where your mission will resonate with your target audience the most. That is why platform selection is more than just tactical, it is also a strategic approach.

For B2B marketers, **LinkedIn** is the linchpin. With over one billion members and a distinct professional audience, it has become the main platform for B2B thinking leaders and a lead generator. To put things in an order of magnitude, **LinkedIn** produces 80% of all B2B leads. https://www.linkedin.com/pulse/linkedin-generates-80-b2b-leads-making-primary-platform-kqd6f/

However, **LinkedIn** is not the only place where operations are being conducted. Email is also essential for developing prospects and delivering restricted speech to end-buyers directly to their personal information systems. Industrial blogs and professional communities offer visibility and deep-dive content. Additionally, search engine optimisation ensures visibility at the most crucial decision-making times.

More importantly, each channel is different. What works on **LinkedIn** won't always work on Instagram or X (Twitter). This is where we believe

AI-powered tools can assist marketers in adjusting to content formats and messaging that fit platform specific behaviour. Webinars and deep-dive reports do well on **LinkedIn**; visual storytelling and culture-focused posts shine on Instagram. Infographics and quick takeaways will perform nicely on Twitter or X. Case studies and how-to guides should live through blogs, newsletters, or gated downloads.

The alignment of content, audience and platform is the point of departure, not a plan. Intent is what makes this a strategy. It's about defining an objective for each channel, understanding what the role is of each of the channels in the journey, and making sure that everything ties in towards a common objective. So, whether you're trying to create awareness, promote consideration, or drive conversion, every message should map to the broader story.

With that under our belts, let's explore how B2B marketers can maximise individual platforms as part of a more focused, strategic mix. I have identified the pros and cons and growing trends of today shaping how marketers can show up and scale effectively on the digital scene.

In the B2B marketing space, **LinkedIn** is the dominant player. It's where serious people go to learn about the industry from their peers and consume long-form content. It has over one billion members and is unrivalled in the professional realm. It is a great tool for demonstrating thought leadership, unveiling new products or services and forging relationships with well-focused content. It's not surprising then that 85% of B2B marketers rate LinkedIn as delivering the best value of any social platform at all. https://contentmarketinginstitute.com/b2b-research/b2b-content-marketing-trends-research

Facebook holds surprising power in B2B when used right, despite **LinkedIn** being the linchpin. It has a reach of more than 3 billion monthly users. Its superiority lies in the precision hit. Facebook's ad platform also offers robust targeting that B2B marketers can apply based on company, role, interest and behaviour, which is great for brand awareness and retargeting campaigns. Recent data shows that 79% of B2B marketers continue to use Facebook. https://contensifyhq.com/blog/b2b-social-media-marketing-trends/ It's a useful platform for mass-awareness campaigns and multi-layered ad funnels.

X, formerly Twitter, remains a real-time source of commentary, industry news, and direct dialogue. Especially if you operate in industries where speed and opinion count, such as tech, finance, and media. About 60% of B2B marketers currently have some form of X in their mix, which they're using for announcements, brand voice and contributing to live event engagement. But recent platform shifts are prompting some to wonder about safety, brand tone and return on engagement: https://www.vogue-business.com/story/technology/as-brands-and-creators-seek-out-x-alternatives-who-wins

Instagram isn't an obvious choice when you think of B2B, but the fact that it's designed around visual storytelling makes it a great way to humanise brands. Marketers are using it to promote company culture, behind-the-scenes looks, team highlights, and even share customer success stories in a more visual and engaging way. Approximately 60% of B2B marketers now use Instagram, sometimes in support of employer brand or culture messaging. https://contensifyhq.com/blog/b2b-social-media-marketing-trends/

In B2B, **YouTube** acts as an education hub. Whether it's explainer videos, product demos, customer case studies, how-to series, or similar, video content serves to make complicated offerings more "digestible" while positioning you as an authority. Approximately 56% of B2B marketers use YouTube as a means to disseminate their content. https://contensifyhq.com/blog/b2b-social-media-marketing-trends/

Then there is **TikTok**. Yet it's hard to ignore its rapid expansion and an audience demographic that is evolving quickly, causing many in the B2B world to view the platform as an experimental medium. Inevitably, some brands do well on TikTok, where short-form video content is king and a creative outlet, mixing humour with authenticity and informal thought leadership. With user growth doubling between 2020 and 2022, it's drawing extra attention from way ahead-of-the-curve teams. https://contensifyhq.com/blog/b2b-social-media-marketing-trends/

Instead of chasing every single platform, savvy marketers follow and measure their audiences' behaviours and prioritise depth over breadth. It's not about spreading yourself thin across the channels but deepening where you matter most to your brand and buyer.

Leveraging AI to Maximise Social Media Interactions

AI is the key, shiny new object when it comes to optimising social media strategies today. This is not merely about automation. It is about amplification. B2B marketers are embracing AI to schedule content more efficiently, get insights more quickly, and engage with the audience in more specific and relevant ways.

Posting schedules can be managed on many accounts in one place with tools like **Sprout Social**. The platforms are increasingly applying machine learning algorithms to recommend the best time to post, evaluate user engagement patterns and suggest changes in real-time. That's why brands can catch shoppers' attention without degrading quality.

Outside of automation, AI is also improving content creation. Tools like **Buffer** and **Jasper** work wonders for creating headlines, social copy, and visuals. They enable teams to maintain a consistent drumbeat of topical, engaging posts, that don't require more time-consuming strategic thinking. Tools like **Quid** and **Brandwatch** are also making it easier to conduct sentiment analysis and social listening to ensure that marketers are ahead of the game in terms of brand perception and audience mood.

93% of marketers use AI to generate content faster, 81% to get insights quicker, and 90% to make faster decisions in real-time (Reported by SurveyMonkey). https://www.surveymonkey.com/mp/ai-marketing-statistics/

Meanwhile, Martech tells us that 76% said they use AI for content generation, 71% for coming up with new creative ideas, and 63% for surveying the market. https://martech.org/ai-and-marketing-what-the-stats-show/

AI in personalisation is particularly powerful. With intelligent post recommendations and custom email sequencing, AI is helping us send the perfect message to the perfect person at the perfect time. That even refers to chatbots on social channels, which are now fielding basic queries, triaging sales questions, and providing around-the-clock presence.

Efficiency is not the only aim. It is resonance. With the use of AI, B2B marketers can become more agile as they hone strategy, gain speed to insight, and stay relevant across platforms that change quickly. The product, in turn, is a more real-time, data-driven, human-to-human social engagement strategy.

Content Syndication and Online Advertising

Content syndication and digital advertising are two of the most potent weapons in the B2B marketer's arsenal. Strategically employed, they can significantly improve visibility, attract qualified traffic, and turn prospects into leads. Both have been turbocharged by artificial intelligence, which provides precision, scale and the ability to adapt in real-time.

Content syndication widens the reach of thought leadership, product messaging, or insights by pushing it through third-party channels. That enables brands to access customers they wouldn't be able to touch using owned channels exclusively. This is where AI helps, surfacing high-fit platforms and suggesting the most efficient formats and timings. Through behavioural data and predictive modelling, marketers can adjust their syndication efforts to meet buyers where they are, not where they might be.

For example, AI-powered programs can evaluate how previous syndication campaigns have performed and recommend which platforms will most likely produce quality leads. The era of "spray and pray" is over. Smart syndication is all about relevance and ROI. Over 80 per cent of global marketers are already integrating machine-learning capabilities into existing marketing strategies. https://www.typeface.ai/blog/content-marketing-statistics

Channel strategy isn't media planning, it's behaviour mapping. From the advertising perspective, AI has made programmatic media buying into something completely different. Advertisers can place, tweak, and optimise ads in real-time, depending on who is looking at them and how they're interacting. Programmatic tech, in a nutshell, enables advertisers to bid in real-time to target audiences based on their behaviour, delivering content to the most likely converters. This optimises media spending and reinforces campaign performance.

This article from **Forbes** describes the way AI is changing digital advertising, including its effects on click-through rates, conversion, and campaign agility. https://www.forbes.com/councils/forbesagencycouncil/2024/04/09/how-ai-is-revolutionizing-digital-advertising-in-2024/

Customisation is also a key benefit of AI in marketing. Creativity can now be adjusted for various personas, industries, and sometimes even individual prospects by leveraging firmographic and behavioural data. Dynamic creative optimisation tools tailor the visuals and messaging based on who is watching the ad, which helps make them more relevant and more effective.

Campaign monitoring is another area on which AI is leaned. Performance dashboards on platforms such as **Brandwatch** enable marketers to quickly see what's working and, just as importantly, what isn't so they can pivot and A/B test in real-time. These resources extend beyond vanity metrics to provide true insight into campaign engagement, channel contribution, and sales impact.

According to **Statista**, the use of AI in content creation is increasing at a fast pace. AI use in marketing: Over 20% of American content creators have adopted AI to generate videos, images or written content. https://www.statista.com/topics/5017/ai-use-in-marketing/

With that growing level of automation and intelligence, we enter a world where content syndication and advertising go well beyond simply driving traffic. They turn into super-tuned demand generation machines that run 24/7 to attract, qualify, and convert.

Marketers who win in these channels don't stop at paid reach. They are investing in data-led creative, channel specialism and technology that can scale to produce personalised experiences. AI doesn't replace the marketer but rather allows them to stretch further, focus sharper, and play to the strengths of serving up the perfect message to the customer at exactly the right moment.

How to Combine Online and Offline Channels

Another factor of a successful B2B channel strategy is the need to blend online and offline experiences. In a world where consumers are fluidly navigating digital and physical landscapes, buyers now often switch back and forth. A conversation initiated at a trade show may result in a webinar download, followed by a series of nurture emails and, ultimately, a meeting with sales. Marketers need to make sure their offline and online strategies sing from the same hymn sheet if consumer confidence and expected brand experience across these touchpoints are to grow. https://www.quadigy.com/blog/cross-channel-marketing-integrating-offline-and-online-in-b2b/

Integration is more than mere presence. It is about cohesion. When offline and online tactics are working together, and the messages make sense, when your systems can talk with each other and your teams work hand in hand across all these touchpoints, then the experience seems entirely connected. This adds credibility to your brand, enhances engagement, and drives conversion.

Take combined campaigns, for instance. These are campaigns that combine print, email, social media and events into one seamless experience. A product launch could involve direct mail teasers, a digital ad campaign, a webinar led by subject matter experts, and follow-up emails with access to downloadable assets. Each element boosts another part in a snowballing effect that no single channel could have delivered on its own.

Event marketing acts as another great connector between offline and online. Trade shows and conferences remain as important as ever for networkers in the B2B arena, but digital amplification can multiply their effectiveness. Now, marketers livestream their company's online sessions, build custom landing pages and serve retargeting ads to re-engage attendees post-event. This would take a three-day event and turn it into a three-month lead generation machine.

Even junk mail has made a comeback. It's now being combined with digital calls to action (e.g., personalised URLs, QR codes) that urge the recipient to access rich online content. That not only quantifies performance but also connects the human touch of tactile engagement with digital experiences. Here is an Oro Inc. case study demonstrating how QR-enabled mailers drove better engagement and conversion. https://oroinc.com/orocrm/blog/5-great-examples-integrating-online-offline-marketing/

When you combine offline and online B2B marketing tactics, you see improved customer engagement and greater performance across the board (DemandLab) https://www.demandlab.com/resources/blog/why-integrating-offline-and-online-marketing-is-essential-for-todays-brands/
Buyers expect consistency at every touchpoint, from reading white papers to attending webinars to shaking your hand at a trade show.

Measuring Performance for Continuous Optimisation

Measurement is the last piece of the equation. Multi-touch attribution tools can now follow and make links between offline and online interactions, enabling marketers to understand how different pieces of content and touchpoints contribute to a deal. I found this study through ResearchGate, and it explains how this integration helps businesses better understand the strategic impact of their marketing investments. https://www.researchgate.net/publication/382580045_Research_on_the_Integration_of_Online_and_Offline_Channels_in_Marketing

The lesson is clear. This doesn't mean that you should just integrate your offline and online content for its own sake. It's about developing an integrated strategy in which all channels have their place, all interactions are deliberate, and every moment contributes toward trust and action.

When done well, integration is more than just achieving operational efficiencies. It enhances the buying experience. It provides marketing and sales teams with a common story. Best of all, it gets results you can feel, measure and replicate.

It's measurement that makes it all come together. Without it, channel strategy is based on hunches. With it, it's a loop of insight and optimisation. B2B marketers, of course, have a longer view than a glance at clicks of impressions when it comes to measuring performance. It's all about understanding what activities affect the pipeline, accelerating decision making and closing deals.

The one thing we should all do before taking a step forward is choose the right key performance indicators. Website traffic shows reach. The 'efficiency' of a sales team is determined by lead conversion rate. Customer acquisition cost is a measure of efficiency, and customer lifetime value is the long-term impact. Click-through rate and bounce rate are good engagement metrics that show how well your audience is engaging with your content. Individually, no one metric tells the entire story, but together, they add up to a picture of what's working and what isn't.

AI brings a second layer of intelligence. It will do more than measure, it will predict. Predictive analytics enables marketers to make predictions about their campaign outcomes and adapt in real-time. Machine learning-powered lead scoring models can identify the ones that are most likely to become customers, driving sales reps to follow up with those leads first. AI can also segment customers more accurately, revealing patterns that might escape manual analysis.

MNTN reports that AI tools are increasingly used to track customer behaviour, analyse engagement patterns, and even predict future marketing outcomes: https://mountain.com/blog/b2b-performance-marketing/

Multi-touch attribution models are a must to join the dots throughout the buyer journey. The purpose of these models is to assist marketers as they assign value across channels from initial awareness through the end conversion. With linear models all touchpoints are considered equal. More recent interactions are given more weight in the time-decay models. U-shape models emphasise the first and last touches. The proper attribution model is going to vary based on your objectives and the length and complexity of your sales cycle.

Bounteous here emphasises the importance of attribution in any modern measurement strategy: https://www.bounteous.com/insights/2024/10/01/enabling-b2b-performance-optimization-through-measurement-strategy/

Optimisation follows insight. It is not something we do one time. It's a rhythm of testing, experimenting and refining. A/B testing allows for experimentation with different headlines, layouts, or call-to-action variations. Benchmarking gives you perspective on your metrics. Feedback loops, particularly those including feedback from customers or even the sales teams, introduce real-world nuance in the optimisation.

The best marketers at measurement are those who use data as a compass, not a scoreboard. They use it to find their way, not to determine what is what (if you get my meaning). They try to find the signals amid the noise of the numbers. They construct tactics that change in response to what buyers actually do, not what marketers wish they would do.

According to Intentsify, tracking the right metrics does more than validate marketing spend. It aligns marketing and sales forces and engenders a collaborative focus on revenue as a responsibility. https://intentsify.io/blog/b2b-marketing-metrics/

After all, performance measurement is the way you keep your channel strategy alive. It allows you to ask the tough questions, to challenge your assumptions, and to outpace change. It provides the confidence to invest, the clarity to prioritise, and the insight to scale what works.

Yes, you can leverage AI and attribution models and be dedicated to continuous improvement, but eventually, measurement is the engine that powers all other components of your strategy toward success.

Summary:

Developing an effective B2B channel strategy necessitates a comprehensive understanding of the omnichannel ecosystem, meticulous selection and integration of marketing channels, and the utilisation of advanced technologies like AI.

- Understand the Omni-Channel Ecosystem
- Evaluate and select the appropriate channels
- Integrate offline and online channels
- Leverage AI for optimisation
- Continuously measure performance.

Case Study: Versa Marketing, Levanta and Grace & Stella

This case study underscores the importance of a well-executed B2B channel strategy. It is the collaboration between Versa Marketing, Levanta and Grace & Stella, facilitated through Amazon's affiliate program. This partnership exemplifies how integrating affiliate marketing can significantly enhance sales and brand visibility. https://www.versamarketinginc.com/case_studies/amazon-affiliate-program-case-study-levanta/

13. Technology and Automation as a Growth Engine

"All the different software in your Martech stack grew in complexity as a function of this inherent complexity in modern marketing."

Scott Brinker, VP Platform Ecosystem at HubSpot and
Editor at chiefmartec.com

The B2B marketing ecosystem has changed significantly due to increased technology and automation. Historical methods of marketing, which had previously been informed by manual effort and limited targeting, have bloomed into machine-learning supported, data-informed campaigns fed by automation and AI. This is no longer just a trend but a paradigm shift in how companies interact with customers, run their operations and scale their revenues.

In this chapter, we will go deep into discussing the key building blocks of modern marketing technology (Martech) stacks, comprising AI in lead nurturing and scoring, the barriers to success for enterprises who are intent on adopting these technologies, and the latest trends in B2B marketing that are shaping your future practise of marketing. As a company, it is important that you adopt these technological improvements as a regular business practice to remain competitive in such a digitalised world where businesses are moving increasingly to digital environments.

Technology and Its Effects on Modern B2B Marketing

There have been some critical innovations in B2B marketing for technology:

Improved Efficacy and Efficiency

Automation tools have trimmed the fat from monotonous processes, allowing the marketing team to turn their attention to more strategic initiatives. For example, the automation of email campaigns, social media scheduling and data analysis workloads ease the human effort and dramatically lowers the chances of bloopers.

A report from **McKinsey & Company** specifies that AI-powered marketing can improve efficiency by 30% and also makes clear the influence of automation on business expansion. https://www.mckinsey.com/capabilities/growth-marketing-and-sales/our-insights/ai-powered-marketing-and-sales-reach-new-heights-with-generative-ai

Data-Driven Decision Making

With the rise of sophisticated analytics software, marketers are able to collect and evaluate an increasing quantity of data, making their decision-making much more informed. There is a lot of focus on data-driven communication, which allows targeted marketing, tracking its performance and achieving optimal marketing strategies.

Martech utilisation has dropped off to 33%, the **Gartner** 2023 Marketing Technology Survey reveals, which has introduced risk and pitted CMOs with a decision to make. CMOs must balance AI pursuits with the need to maximise returns on existing investments. https://www.gartner.com/en/marketing/topics/marketing-technology

Personalisation and CX

Tools with AI can be used to personalise content and experiences to maximise customer engagement and satisfaction. Through customer behaviour and preference analysis, organisations can personalise their messaging.

According to a recent **Intuit Mailchimp** survey, more than two-thirds of marketers say increasing the use of marketing automation is important to bring in and maintain customers. https://mailchimp.com/intuit-mailchimp-finds-vast-majority-of-smb-marketers-are-bought-into-ai/

Scalability and Flexibility

Technology allows businesses to automate and scale marketing with ease.

A single experience can be generated and maintained across multiple campaigns on different channels without a requirement that resources scale proportionally.

Gartner forecasts a substantial rise in network automation by 2026, with 30% of enterprises automating more than half of their network activities, compared to less than 10% in mid-2023. https://www.gartner.com/en/newsroom/press-releases/2024-09-18-gartner-says-30-percent-of-enterprises-will-automate-more-than-half-of-their-network-activities-by-2026

Core Martech Stack Components

An effective Martech stack is the heart of sustainable B2B marketing campaigns. These are the tools that allow businesses to automate tasks, make data-driven decisions, and provide personalised customer interactions at scale. Though Martech applications can differ based on business requirements, four fundamental elements are essential to any successful stack.

1. Customer Relationship Management (CRM) Systems

A CRM system is vital to managing customer interactions, sales activity tracking and facilitating better communication between marketing and sales teams. It's a place to store all prospect and customer data in a central location to make sure marketing and sales are aligned.

According to **Salesforce** research, there is a 29% rise in lead quality and a 34% rise in productivity for companies using CRM. https://www.salesforce.com/ap/hub/business/how-salesforce-improves-business-performance/

2. Marketing Automation Platforms (MAPs)

Marketing automation tools make it easier to manage repeated marketing tasks (email, social, etc.), as well as track the several stages of a marketing funnel. They enable efficiency and the ability for marketers to provide prospects with meaningful messaging.

According to Invesp, businesses using marketing automation platforms experience an average of 14.5% increase in sales productivity and a 12.2%

reduction in marketing overhead costs. https://www.invespcro.com/blog/marketing-automation/

3. Customer Data Platforms (CDPs)

Unlike CRM platforms which are more sales or account focused, CDPs consolidate customer data from different sources to create a single profile of a customer for marketers so they can offer more personalised marketing across all channels.

A study found that companies using CDPs for real-time data management achieve a 25% increase in marketing ROI. https://useinsider.com/benefits-of-a-customer-data-platform/

4. Analytics and Business Intelligence (BI) Tools

Data-driven decision-making is at the heart of modern B2B marketing. Analytics and BI tools provide real-time insights into campaign performance, customer engagement, and sales effectiveness, enabling continuous optimisation.

Research by **Forbes** found that companies using advanced analytics improve their decision-making by 50% and see a 15% improvement in efficiency. https://www.forbes.com/councils/forbestechcouncil/2024/10/23/leading-with-data-bi-and-analytics-for-business-success/

Integrating the Core Martech Stack for Success

A complete Martech stack facilitates the more effective functioning between the marketing and sales teams, ensures stronger, more personalised customer experiences, and allows more strategic decisions using data to drive better business results. But it is important that these tools have interoperability to discourage data silos and lack of efficiency.

As AI and automated solutions develop and businesses look to apply them to their Martech strategy, it will become increasingly important to work with a Martech stack that has a degree of flexibility to scale and adapt to any new marketing innovations that become available.

Marketing Automation and Lead Nurturing

Marketing automation is one of the cornerstones of modern B2B marketing, and it allows companies to reach potential customers at scale, nurture leads efficiently, and deliver those leads to the sales team. Automation without intent is just noise at scale. Smart marketers use it to enhance the experience, not accelerate interruption.

AI-enabled automation now drives the entire life cycle from email sequences and personalised content to predictive lead scoring and chatbot conversations - all of which are designed to make marketing work smarter and improve the buying audience experience.

As the B2B purchase process becomes more complicated, involving multiple stages, businesses need to employ automation to sustain a uniform and personalised appearance across numerous touchpoints.

Tailored Multi-Channel Engagement

Marketing automation enables companies to personalise their communications depending on a leads past interactions with your company, industry, job title, and stage in the buying cycle. All this is personalised at scale so that prospects get content and messaging that resonates with them and drives conversion.

A study by **Salesforce** found that sixty-six percent of B2B buyers expect brands to personalise engagement based on past interactions, underscoring the need for automation in lead nurturing.
https://www.salesforce.com/resources/research-reports/state-of-the-connected-customer/

Lead Scoring and Qualification Automation

Manual lead scoring and qualification are slow and subjective. Machine learning and predictive analytics are leveraged to evaluate if a lead is likely or not to convert based on its behaviours, engagement and firmographics.

Drip Campaigns and Behavioural Triggering

With marketing automation, businesses can send personalised, scheduled messages to send them down their pipeline with drip campaigns. These dynamic sequences help ensure that leads are getting the right message at the right time, allowing them to stay engaged as they move through the buying process. Behavioural based triggers (i.e., download a white paper or register for a webinar) can change the messaging automatically based on prospect behaviour.

HubSpot research shows that businesses using automated email workflows generate a fourfold increase in qualified leads compared to those using manual follow-ups. https://blog.hubspot.com/marketing/email-marketing-trends

Automation As The Basis Of Scalable B2B Marketing

Not only does marketing automation allow businesses to develop meaningful interactions with prospects, but it also helps them streamline resources and ensure better lead management. As AI progresses, organisations need to incorporate automation into their marketing just to keep up. Personalised engagement, predictive lead scoring and behaviour-driven campaigns are no longer nice to have; they are a must for B2B success in today's market.

The Role of AI in Predictive Analytics and Personalisation

AI is revolutionising B2B marketing by enhancing predictive analytics and enabling real-time personalisation at scale. In the past, traditional marketing was focused on retrospective data and broad audience segments, while now, AI empowers companies to make sense of massive real-time streams of behaviour and intent and predict what the consumer needs next. This adjustment helps marketers provide custom-tailored messaging, boost conversion rates, and foster greater customer rapport.

How B2B Marketing is Improved through AI-Driven Predictive Analytics

Predicting Buyer Intent And Behaviour

Marketing can use predictive analytics to find out when customers are most likely to buy by analysing online behaviour, consumption patterns, and engagement. AI-powered models yield lead scores for the probability of closing, so marketing and sales can focus on higher-value accounts.

A study by **Forrester** found that businesses using predictive analytics for lead scoring experience a thirty-six percent higher conversion rate compared to those relying on traditional scoring methods. https://get.nice.com/rpcs-forrester-automation?

Real-Time Hyper-Personalisation

Personalisation has become about more than just inserting a prospect's first name into an email. AI processes customer data across touchpoints in order to offer content, product recommendations, and campaign approaches that are in line with individual preferences and behaviour.

McKinsey research found that AI-driven personalisation can increase revenue by ten to fifteen percent and improve marketing ROI by up to thirty percent. https://www.mckinsey.com/featured-insights/mckinsey-explainers/what-is-personalization

Dynamic Content Adaptation

Businesses can dynamically alter content based on AI-driven customer engagement data. The content on the website, ad creatives and email sequences can be adjusted based on the choices of the user, so they receive a personalised experience that is tailored specifically for them.

LinkedIn reports that seventy percent of B2B decision-makers engage with personalised content before making a purchase decision, emphasising the role of AI in driving engagement. https://www.linkedin.com/pulse/how-b2b-buyers-research-online-before-making-purchase-tfa1f/

The Future of B2B Marketing Intelligence is AI

Artificial intelligence based predictive analytics and personalisation are transforming how businesses engage with both prospects and customers. Businesses that adopt these technologies have a competitive edge by predicting buyer demands, fine-tuning their marketing activities, and providing hyper-personalised experiences. Whether or not new and existing technologies expand upon it to be a 'must have' in the B2B Marketer's toolkit is, of course, yet to come.

The Challenges of Merging Marketing Automation and AI

Though marketing automation and AI-based approaches hold real merit, implementing such technologies remains an issue for many enterprises. Typical issues are siloed data, automation fatigue, concerns around compliance, and resistance to change among marketing and sales teams. Addressing these issues is essential to optimise the ROI from AI-based marketing solutions.

Key Problems and Solutions

Breaking down silos for integrated data interflow.

Clean, organised data is essential for marketing automation tools to work as they should. However, fragmented data across numerous systems can pose a problem to many businesses, making it hard to establish a single customer view.

Gartner research reveals that 81% of companies have identified data silos as a considerable hurdle to achieving marketing automation success. https://www.gartner.com/smarterwithgartner/data-sharing-is-a-business-necessity-to-accelerate-digital-business

Rather than throwing data at your customer journey and hoping that it sticks, businesses need to invest in a Customer Data Platform (CDP) to bring disparate data together and lift it to de-facto standards. Establishing cross-functional integration between marketing, sales, and IT teams improves data governance and reduces discrepancies.

Preventing Automation Fatigue and Promoting a Human-Centric Interaction

When you rely on automation too much, you end up with impersonal, copy-paste sounding emails that don't resonate with prospects. Automation adds efficiency, but businesses need to combine it with human-led engagement methods.

HubSpot shows that 68% of consumers say they are willing to pay more for products and services from a brand known to offer good customer service. experiences. https://cdn2.hubspot.net/hubfs/2771217/Content/2018%20Customer%20Service%20Expectations%20Gladly.pdf

Leverage automation with AI for repetitive work while handing off your touchpoints, such as sales calls and personalised emails, to humans. Utilise AI-enabled chatbots to hand over difficult questions to humans instead of only automated responses.

Navigating Compliance and Ethical Considerations in AI Marketing

As AI-based marketing takes off, so do concerns over data privacy, compliance, and appropriate use of customer data. Laws like GDPR and CCPA have stringent rules for how companies gather and use personal data.

According to **Salesforce**, 56% of customers are glad to give their personal information in return for better service. https://www.salesforce.com/form/pdf/state-of-the-connected-customer-2nd-edition/

Businesses must prioritise transparent data collection practices, ensuring that AI models comply with GDPR, CCPA, and other regulations. AI ethics guidelines need to be created to avoid bias in predictive analytics and targeted marketing.

Overcoming Barriers to AI-Driven Marketing Success

The power of AI and automation in B2B marketing is apparent, but companies need to overcome common obstacles to reap the benefits. With the removal of silos, the right balance between automation and the human touch, and abiding by data protection laws all in check, companies can create a marketing strategy that is not just effective but ethical too. AI is not going away, but businesses need to approach how they think about leveraging it in marketing if they want to find lasting success.

Key Trends Shaping the Future of B2B Marketing

Artificial intelligence (AI) and automation are proliferating in B2B marketing at an increasing pace. Businesses that see the writing on the wall (or at least on the screen) can adapt, get ahead and compete, while those that don't evolve find themselves increasingly and hopelessly lagging. As AI continues to evolve, B2B marketers should take a progressive approach to unlocking the full power of automation and machine learning.

AI-Powered Predictive Analytics Will Guide Decision Making

Predictive analytics will be even more advanced, helping marketers offer product suggestions even more accurately. Artificial intelligence models will analyse real-time behavioural information, market trends and historical learnings to forecast purchasing patterns and suggest the most effective engagement tactics. https://www.forrester.com/press-newsroom/forrester-predictions-2025-b2b-marketing-sales/

Conversational AI and Voice Search Optimisation Will Expand

Voice search and conversational AI are transforming the way companies communicate with their prospects and customers. Chatbots and voice-based virtual assistants driven by AI will be a key part of lead nurturing and customer service in a natural language in real-time. https://searchengineland.com/search-engine-traffic-2026-prediction-437650

The Norm Will Be Hyper-Personalisation

As AI progresses, personalisation will shift away from static audience segmenting and towards dynamic, personalised experiences in real-time. Businesses will utilise AI-generated insights to personalise content, email campaigns, and web experiences on the fly according to a user's actions, clicks and preferences.

Ethical AI and Accountable Automation are Going to Matter More Than Ever

When companies become more ambitious in their use of AI in marketing ethical questions around data privacy and bias in algorithms, understanding AI will become more and more crucial. Greater regulatory requirements will be brought in by governing bodies, with ethical AI adoption being a business need.

AI is Here to Stay

AI, automation, and data-based decision-making will define the future of B2B marketing. Companies that adopt these trends will drive more relevant and engaging customer experiences while optimising their marketing effectiveness. But AI is no substitute for human creativity and strategic thought. Effective marketing in the age of AI strikes a balance between automation and the human touch to make sure technology empowers, not replaces, the discerning customer to engage with a brand.

Businesses that challenge the status quo invest in AI thoughtfully, and smoothly adapt to automation will emerge as early leaders in the next era of B2B marketing.

Responsible AI: Building Ethical Foundations for the Future

With increasing capabilities, there are also increasing ethical challenges regarding their use. In the world of B2B marketing, this has been replaced by technology to segment buyers, predict behaviour, create content and automate decision-making at scale. But such power comes with great responsibility.

Abuse of AI can give rise to mistrust and resentment among a consumer base or damage your reputation. Marketers must, therefore, take the lead with ethical intent, not just technical excellence.

Here are some key principles for responsible AI in marketing:

Transparency: If AI is driving decisions, what content is displayed, what score was assigned to a lead, what email was sent, share it openly internally and also where it makes sense with the customer.

Consent and data respect: Ensure that permissioned data sources are privacy-compliant and not inferred irrespectively of ethical boundaries.

Bias monitoring: Programs amplify historical bias, discriminating by industry, region or job title without intending to. It is important to conduct routine monitoring of outputs.

Human oversight: AI should guide, not make decisions. Accountable humans should make the final calls about strategy, messaging and segmentation.

Regulation will come. However, ethical marketing cannot wait for the law. If brands apply AI responsibly, transparently, and for the benefit of customers, they'll gain trust and competitive advantage.

Responsible AI isn't just about compliance. It's a brand issue. In the grand scheme of things, your ethical decisions might be a more significant metric than your click-through rate.

Summary:
Technology and automation streamline complex marketing processes, enabling scalability and precision. Martech tools are indispensable for modern B2B marketing.

- Invest in marketing technology that aligns with your goals.
- Automate repetitive tasks to focus on strategic activities.
- Continuously evaluate and adapt your Martech stack.

Case Study: Marketo's Automation Success
Marketo revolutionised B2B marketing by automating repetitive tasks like email campaigns and lead scoring. This allowed teams to focus on strategic activities and improved lead nurturing. Marketo's ability to streamline marketing efforts became a benchmark for success in marketing automation.
https://growthnatives.com/blogs/marketo/marketing-automation-insights-marketo-business-experts/

14: From Gut Instinct to Data Confidence

"Without data, you're just another person with an opinion."

Avinash Kaushik, Digital Marketing Evangelist

In marketing, strategic decision-making is a strategy based on data. It also allows marketers to refine campaigns, gain deeper insight into buyer behaviour, and increase ROI by being more targeted. Through the smart use of data, B2B marketers can provide highly personalised, targeted campaigns that stick with their audiences, get more dollars out of the budget, and make the case to stakeholders. In an AI focused, predictive analysis, customer data platform driven world, the ability to leverage and interpret data is no longer a 'nice to have'; it's a requirement.

This chapter examines how data has revolutionised the way B2B marketers can evaluate the effectiveness of their efforts by covering the types of data that are now available, key metrics that should be followed, essential tools and best practices, predictive analytics, and overcoming common challenges.

AI and The Move to Data-Driven Decision Making

The role of data in B2B marketing has undergone a dramatic shift in the past ten years. Many traditional marketing efforts were based on gut feelings, what has worked in the past, and the mass marketing segmentation used to make marketing decisions.

The data confidence maturity model:

1. **Reactive (gut feel, unstructured data)**

2. **Reporting (basic dashboards)**
3. **Insight-driven (actionable metrics)**
4. **Predictive (forecasting, modelling)**

The proliferation of AI and advanced analytics is changing how data is being gathered, processed, and used by businesses, enabling better, predictive, and more personalised marketing campaigns.

The increasing amount of data produced by digital engagement offers marketers both opportunities and challenges. In addition, access to huge amounts of data gives companies a better view of customer behaviour, but companies are more often than not challenged by information overload, data quality, and insights. The problem isn't a lack of data. It's a lack of interpretation. The ability to cut through this complexity and convert data into strategic 'intelligence' is now a critical factor in the success of B2B marketing.

A report by **McKinsey & Company** demonstrates the potential of data-driven decision-making, especially in marketing and sales. Their research into data-led commercial performance demonstrates that businesses that use advanced analytics in decision-making are driving greater returns on investment from the many different touchpoints they have with their customers and in multiple channels. https://www.mckinsey.com/capabilities/growth-marketing-and-sales/our-insights/insights-to-impact-creating-and-sustaining-data-driven-commercial-growth

Furthermore, data-driven companies are much more likely to win new customers and be profitable. https://exclaimer.com/blog/maximise-your-marketing-roi-with-data-driven-strategies/

Now, artificial intelligence doesn't just mean the automation of monotonous, repetitive marketing functions. It's changing the way companies think about customer engagement, lead qualification and campaign performance analysis. AI powered predictive analytics can return buying signals before prospects take any active interest, helping you interact with customers at the right time. Machine learning programmes allow marketers to adjust messages in real-time to make them more specific, increasing the conversion rate of messages and reducing wasted marketing dollars.

With the continued rise of AI-led analytics, businesses that take a data-first approach will be more likely to have the tools to inform decision-making based on a data-first marketing approach, with the ability to pivot and cater to the needs of the customer, with the ultimate goal to drive marketing performance optimisation.

Key Types of Data in B2B Marketing

Making the best decision possible is based on the ability to gather the pertinent data and use that data for analysis effectively. Businesses dependent on fractured or obsolete information will assume unknown strategic liabilities, leading to inefficiencies and missed opportunities. AI is also refining data collection, and that means insights can be not only accurate but timely and actionable as well.

B2B marketers use different types of data to inform decisions. They all have a unique role to play in interpreting customer behaviour, optimising engagement strategies, and boosting overall marketing performance.

Demographic Data

Demographic information gives baseline information about people in target organisations. This includes job titles, industries, company size, geographic location, and professional experience. Knowing who the people in the company are making the decisions and who can play a key role helps marketers make decisions about what to say and who to say it to.

Artificial intelligence improves demographic data analysis by enriching contact profiles with data from databases. Artificial intelligence powered tools such as **ZoomInfo** and **Clearbit** are capable of pulling live data, which means you are not dependent on outdated contact lists.

Firmographic Data

Firmographic data tells you more about a company rather than about the individuals within it. That information contains size, revenue, industry and growth. This kind of data is necessary for account-based marketing, as it helps companies segment targeted accounts by key firmographic signals.

AI-driven customer data platforms, including **Segment** and **BlueConic**, unify firmographic data across multiple sources to give companies a full picture of target accounts. Powered by AI-based firmographic intelligence, marketers can focus on high-value accounts and personalise outreach with more precision.

Behavioural Data

Behavioural data monitors how prospects and customers engage with a company's marketing assets, website and digital content. This can range from page views and email activity to content downloads, webinar attendance, and social media engagement. Analysing behavioural data helps companies identify which prospects are most engaged and which are most likely to convert.

AI enhances behaviour analysis, finding trends in how customers are interacting. With AI-driven marketing automation technologies like **HubSpot** and **Marketo**, leads are rated using an engagement score rating, so salespeople know where to focus efforts.

Transactional Data

Transactional data includes records of past purchases, subscription details, contract values, and renewal history. Understanding how your customers buy allows businesses to identify upsell and cross-sell opportunities, such as predicting churn, and optimise your pricing strategy.

Predictive analytics Workbench tools with AI, such as **Salesforce Einstein** and **Clari**, leverage historical transactional data, making it possible to forecast customer lifetime value and future buying behaviours. Such understanding is valuable information for businesses to optimise sales and retention strategies.

Intent Data

Intent data indicates which prospects are actively searching for solutions that they need, which, in return, helps to understand more about where they already are in the buying process. This data is captured from online behaviours like product searches, content consumption, and researching competitors.

AI powered intent data platforms such as **Bombora** and **6sense** dissect online behaviour to identify companies that are exhibiting buying signals. With the inclusion of intent data as part of their marketing arsenal, companies can reach out to prospective buyers just in time, thus increasing the possibility of a conversion.

Engagement Data

Engagement data is the measure of how customers and prospects engage with marketing campaigns, sales outreach and customer support. This includes number of emails opened, time until reply, how likely they are to attend a webinar, and how often they download files.

Artificial Intelligence powered customer engagement platforms like **Intercom** and **Salesloft (Drift)** offer real-time visibility into engagement patterns to help companies optimise their approach to communication. AI can further personalise engagement using an individual's past interactions, thus improving relevance and response rates.

Customer Sentiment Data

Customer sentiment analysis collects feedback from reviews, surveys and social media to learn how customers feel about a brand, product or service. Sentiment analysis means businesses can understand how the brand is perceived and can make necessary changes, where necessary, to prevent this from affecting customer relationships.

AI-driven sentiment analysis tools like **MonkeyLearn** and **Lexalytics** feed through troves of customer feedback, pulling out trends in satisfaction and dissatisfaction. These revelations assist companies in tweaking the messaging better for customer experience and the image of the brand.

Predictive Data

Predictive data uses machine learning, and previous behaviour patterns are used to predict what customers will do next. This applies to predictive lead scoring, churn forecast and demand forecast.

AI-based predictive analysis tools such as **Infer** utilise intelligent algorithms to analyse and act upon very large sets of information to uncover patterns that might predict future behaviour. These insights can guide businesses in proactive marketing and sales decisions.

All this information is also incredibly important in determining future B2B marketing strategies. AI also increases the accuracy, relevance, and predictability of data, enabling organisations to optimise marketing in the moment. By leveraging a combination of demographic, firmographic, behavioural, transactional, intent, engagement, sentiment, and predictive data, businesses can create more targeted, effective, and data-driven campaigns.

Metrics, KPIs, and AI-Driven Insights

A successful B2B marketing strategy needs a set of key performance indicators (KPIs) that work together to indicate how effective a campaign is, how actively customers are engaged, and what the ROI is. Key performance indicators like conversion rates and lead generation are still crucial, but AI is changing how business owners monitor, measure and interpret marketing performance. AI driven analytics help in forecasting decisions in real-time and choosing strategy in a dynamic mode and not in a look back model.

KPIs in B2B Marketing

Most B2B marketers use several KPIs to measure performance and find the best ways to improve their marketing. Metrics that matter the most directly support business goals, offering actionable insights into customer behaviour and marketing effectiveness.

Metrics for Lead Generation and Qualification

Quality leads are a top priority for B2B marketers. Counting the number of leads isn't enough; companies need to evaluate the quality and value of a potential lead.

- **Marketing Qualified Leads (MQLs):** These are candidates who are qualified but not yet ready for the sale; the prospect has shown some kind of interest or intent for a certain way.
- **Sales Qualified Leads (SQLs):** Leads that qualify for sales based on pre-determined criteria and show high purchase propensity.
- **Cost Per Lead (CPL):** It is the ratio of total marketing expenditure to the number of leads acquired.
- **Conversion Rate of Lead to Customer:** Proportion of leads getting converted into paying customers.

AI-enabled lead scoring systems like **Salesforce Einstein** and **HubSpot** AI process behavioural data, engagement history, and firmographic traits to determine which lead candidates are ripe for conversion.

Metrics of Engagement and Content Performance

B2B buyers interact with different digital touchpoints prior to the purchase. Measuring engagement on these channels allows marketers to determine content performance and audience engagement.

- Email open and click-through rates: Measures of how good of a job email campaigns have done in capturing the interests of the recipients.
- Time on Page: Time spent by the visitor-viewer on your web page.
- Bounce Rate: The rate at which visitors leave a site without engaging.
- Attendee Rates for Webinars and Live industry Events: Measures the success of company presenter opportunities.

AI-backed analytics tools like **Google Analytics 4** and **Marketo** Measure monitoring track engagement trends and will inform content optimisation based on user behaviour.

Conversion and Sales Metrics

Tracking conversion rates from initial engagement to revenue generation is essential in evaluating marketing effectiveness.

- **Customer Acquisition Cost (CAC):** The total cost of acquiring a new customer, including marketing and sales expenses.
- **Customer Lifetime Value (CLV):** The projected revenue a customer will

generate over their relationship with a business.

- **Pipeline Velocity:** The speed at which leads progress through the sales funnel.
- **Win Rate:** The percentage of opportunities that convert into closed deals.

Predictive analytics tools such as **Clari** and **6sense** predict sales trends and help pipeline management efforts by identifying which leads are most likely to move towards conversion.

Retention and Customer Satisfaction Metrics

B2B marketing doesn't stop at acquisition. Keeping existing customers as part of your growth strategy.

- **Customer Retention Rate:** The percentage of customers who continue doing business with a company over a specific period.
- **Net Promoter Score (NPS):** Measures customer satisfaction and likelihood of recommending a company.
- **Churn Rate:** The percentage of customers lost within a given period.
- **Upsell and Cross-Sell Rates:** Indicators of additional revenue generated from existing customers.

Sentiment analysis tools like **MonkeyLearn** or **Lexalytics** are AI-enabled and listen to customer feedback, reviews, and social media chats to measure the perception of your brand and customer experience.

AI-Driven Performance Optimisation

AI also boosts marketing analytics by collecting data and finding patterns more quickly, as well as predicting trends.

- **Real-Time Reporting:** AI enables marketers to access real-time campaign performance data, allowing for instant adjustments to improve results.
- **Predictive Analytics:** AI forecasts future marketing outcomes based on historical data, helping businesses anticipate customer behaviour.
- **Automated A/B Testing:** AI analyses multiple campaign variations simultaneously to determine which strategies deliver the best performance.

- **Attribution Modelling:** AI evaluates the impact of each marketing channel on customer conversions, ensuring accurate resource allocation.

Automation and AI providers like **Adobe Sensei** and **Google AI** deliver actionable insights at a speed and accuracy that even the most experienced marketer cannot achieve on instinct alone.

The performance metrics are basic and fundamental indicators for the effectiveness of marketing. However, artificial intelligence is re-inventing methods to analyse and exploit these metrics. By using AI in their analytics, companies can make decisions better and have better campaign performance and higher ROI.

Analytics Tools and Technologies Transforming B2B Marketing

Artificial intelligence and advanced analytics are transforming the way B2B brands gather, interpret, and respond to data. Legacy data analysis practices usually involved static reports, manual analyses, and delayed/actionable insights, leading to that 'I needed to know this yesterday' frustration. Today, analytics tools fuelled by AI allow companies to monitor performance as it happens, uncover new trends as they emerge, and automate complex data analysis.

Choosing the tools that best fit is key to getting the most out of data-driven marketing. Businesses will need to use a mix of CRM platforms, marketing automation software, predictive analytics solutions, and data visualisation tools to understand the big picture for their marketing results.

CRM Platforms

CRM systems keep a record of customer interactions and help businesses capture relationships, automate follow-ups, and personalise engagement. AI-powered CRM tools take this feature a step further by employing machine learning to score leads, foresee customer actions, and suggest the best approaches for customer engagement.

Salesforce Einstein provides AI-powered insights into sales opportunities, helping teams prioritise high-value leads based on past interactions and engagement history.

HubSpot CRM automates lead tracking and customer interactions while offering AI-driven recommendations for outreach and follow-up.

Microsoft Dynamics 365 AI analyses customer data in real-time, helping businesses optimise sales and marketing alignment.

Automation Software for Marketing

Marketing automation is the lifeblood of executing campaigns, nurturing customers, and keeping them engaged. AI has made these platforms more intelligent, with capabilities for dynamic audience segmentation, personalised messaging, and predictive analytics to increase the effectiveness of marketing. Companies that use privileged access automation with AI capabilities can scale engagement to drive best-in-class efficiency and ROI. Here are some providers of marketing automation software:

- **HubSpot Marketing Hub** provides AI-driven email marketing, automated workflows, and predictive lead scoring to personalise outreach and optimise engagement.
- **Marketo Engage** uses AI to refine email marketing, automate lead nurturing, and improve audience segmentation based on real-time behavioural data.
- **Pardot by Salesforce** integrates AI to deliver account-based marketing, predictive lead scoring, and personalised content recommendations.
- **Eloqua by Oracle** leverages AI-powered analytics to enhance lead qualification, campaign orchestration, and marketing personalisation.
- **ActiveCampaign** automates customer journeys with AI-driven messaging, lead scoring, and predictive content recommendations.
- **Autopilot** offers visual marketing automation that enables AI-powered customer journey mapping and multi-channel campaign execution.

Predictive Analytics and AI Based Insights

Predictive Analytics tools use data and machine learning technology to predict customer behaviour, sales trends, and even marketing performance. These smart platforms enable businesses to make the right decisions proactively and invest resources in the most effective ways to get higher conversions.

- **6sense:** AI is used to analyse intent data to predict which accounts are in the market for a company's products and services.
- **Clari:** Clari is focused on helping sales and marketing teams optimise their efforts with AI-based revenue forecasting based on your pipeline.
- **Infer:** Leverages machine learning models to determine lead quality and highlight the most promising prospects.

Real-Time Analytics and Visualisation of Data

Combining this with real-time analytics offers insights into campaign performance, audience interaction and sales pipeline activity on the fly. In conclusion, AI-based data visualisation tools simplify sophisticated sets of data, again enabling companies to discover what trends are actionable and how marketing strategies can be optimised.

- **Google Analytics 4** uses machine learning to provide predictive insights on customer behaviour, traffic patterns, and engagement trends.
- **Tableau by Salesforce** offers AI-powered data visualisation, enabling marketers to create interactive dashboards and uncover insights from large datasets.
- **Power BI** by **Microsoft** integrates AI-driven analytics to help businesses track marketing performance and customer journeys.

AI-Powered Customer Data Platforms (CDPs)

CDPs unify data from multiple sources to create a comprehensive view of each customer. AI-enhanced CDPs help businesses segment audiences more effectively, personalise marketing efforts, and improve customer retention strategies.

- **Segment** consolidates data across different platforms, providing AI-driven insights for personalised marketing campaigns.
- **BlueConic** enables businesses to build unified customer profiles and apply AI-driven predictions to enhance engagement.
- **Lytics** uses machine learning to automate audience segmentation and personalise marketing interactions.

Social Media and Sentiment Analysis Tools

Customer sentiment is core to honing brand messaging and customer engagement tactics. Sentiment analysis tools enabled by AI, meanwhile, allow companies to monitor the public's perception of their brand, enabling them to engage transparently with and even respond to customer feedback as it happens.

- **Brandwatch** uses AI to monitor social media discussions, providing sentiment insights and trend analysis.
- **Sprinklr** analyses customer interactions across multiple digital channels to measure brand sentiment and engagement.
- **Lexalytics** applies natural language processing to assess customer reviews, social media posts, and survey responses.

How AI is Bringing Data Together Across Platforms

One of the key challenges for businesses is to bring data from different platforms together in order to provide an overarching snapshot of how their marketing efforts are performing. Integration solutions can help solve this problem by automating the data extraction and creating consolidated and standardised reporting across tools.

- **Zapier AI** automates data workflows, connecting different marketing and sales platforms without requiring manual intervention.
- **RudderStack** synchronises customer data across platforms, ensuring a consistent and unified data pipeline.
- **Talend Data Fabric** applies AI to data integration, allowing businesses to extract, clean, and manage data from various sources.

Today's B2B marketing technology stack is powered by AI-driven analytics solutions that allow businesses to gather, interpret, and respond to data in real-time. Selecting the right combination of platforms is critical for optimising marketing performance, improving lead conversion rates, and enhancing customer relationships. As AI technology continues to advance, businesses that invest in AI-driven analytics solutions will gain a competitive edge in an increasingly data-centric marketplace.

The Rise of Predictive Analytics and AI in Decision-Making

Traditional B2B marketing strategies relied on historical data and periodic reporting to assess performance and adjust campaigns. Although these approaches delivered insights, they had difficulty predicting customer demand in real-time. Artificial intelligence and predictive analytics have changed how businesses are basing their decision-making on data. Predictive analytics helps marketers understand the patterns of customer behaviour, intent signals and engagement history, allowing them to predict demand, optimise marketing spend and tailor outreach.

What are Predictive Analytics?

Prediction analytics involves applying a range of techniques, including statistical analysis and machine learning, to predict future behaviour based on historical and real-time data. From patterns in an individual's past behaviour, predictive analytics can help businesses uncover high-value leads, optimise sales pipelines and fine-tune strategies around content. Unlike traditional analytics, which focuses on past performance, predictive analytics provides forward-looking insights that drive proactive decision-making.

AI-based predictive analytics systems like **Google Cloud AI** leverage large datasets to create models that predict customer behaviour and suggest strategic marketing initiatives.

Key Applications of Predictive Analytics in B2B Marketing

Predictive Lead Scoring

Predictive lead scoring ranks prospects based on their likelihood to convert into customers. Traditional lead scoring models rely on predefined rules, whereas AI-powered predictive scoring dynamically adjusts rankings based on real-time behavioural data, engagement patterns, and firmographic attributes.

Companies using AI-driven predictive lead scoring can prioritise high-value accounts and improve conversion rates by focusing resources on prospects with the highest likelihood of purchasing.

Sales Forecasting and Pipeline Optimisation

AI powered sales forecasting models use predictive analytics to predict their future revenue, extract sales trends, and manage their pipeline more efficiently. Such tools help companies set achievable goals, distribute resources efficiently and predict potential snags in the sales process.

- **Clari** provides AI-powered revenue forecasting by analysing deal progression and historical win rates.
- **Anaplan** integrates predictive analytics into sales planning to improve quota setting and pipeline management.

With the help of predictive sales forecasting, companies can better connect and coordinate marketing and sales initiatives, helping to drive consistent pipeline performance.

Customer Segmentation with AI

AI-based predictive segmentation allows businesses to group customers according to their common traits, engagement habits and purchase intention. This level of granularity allows for hyper-personalised marketing campaigns that resonate with each segment's unique needs.

Predictive segmentation allows companies to drive engagement by delivering content, offers and outreach based on specific customer needs.

Customer Churn Prediction and Retention

Keeping your existing customers is as important as gaining new ones. Artificial intelligence-based predictive analytics models identify early indicators of customer frustration by evaluating behaviours, support interactions and sentiment data. By identifying at-risk customers, businesses can take proactive steps to improve retention.

- **Totango** monitors customer health scores and predicts churn risks based on engagement history.
- **Gainsight PX** applies AI-driven insights to identify customers likely to disengage and recommends targeted retention strategies.

Businesses which adopt AI-driven churn prediction models can lower customer churn and improve the lifetime value of a customer by proactively addressing pain points, potentially even before they occur.

Predictive Content Recommendations

Content recommendation engines that are powered by AI use data to automatically recommend the right content to customers based on customer preferences and engagement. These are the types of tools that enable companies to succeed with content marketing so that they can provide prospects and customers with the type of information they need when they need it.

The likes of **PathFactory** monitor the behaviour of buyers and suggest content that is relevant to the stage of purchase they're currently in. Fuelled by AI-based content recommendations, businesses can raise content engagement questions and shorten the buying cycle.

The Role of AI in Enhancing Predictive Analytics

AI is a key catalyst for improving the accuracy and actionability of predictive analytics. Machine learning algorithms get better at analysis through repeated refitting to new data points whose insights sharpen with each refinement. The benefits of AI on predictive analytics include:

- Automated Model Training: AI continuously refines predictive models by incorporating new data, ensuring accuracy in forecasting.
- Real-Time Data Processing: AI enables businesses to act on predictive insights immediately rather than waiting for scheduled reports.
- Personalised Recommendations: AI tailors marketing and sales recommendations based on customer behaviour, improving conversion rates.

B2B marketing is finally coming to the party where Predictive Analytics is mission critical for companies to get lead generation, sales forecasting, customer segmentation, retention programs, and content recommendations. By using AI-powered predictive analytics as part of their marketing stack, companies can predict what customers need, tailor how they engage with them and gain a greater return on investment.

Best Practices for Data-Driven Marketing

Creating successful data-driven marketing practices involves systematic data collection, analysis, and usage. Artificial intelligence and predictive analytics allow businesses to create intelligent marketing strategies, but they truly become powerful when data is deployed responsibly, ethically, and with strategic vision. Bad data management can result in incorrect insights, potential risks with compliance, and waste of resources. Best practices for sensitive data-driven operations can help businesses get the most out of their processes while maintaining transparency and responsibility.

It's All About Accuracy And Precision

The decisions marketing makes are only as strong as the data backing up those decisions. False, irrelevant, or old data will tease you with detracting insights. Organisations need to have procedures in place to ensure that the data remain consistent and free of errors.

- Automated Data Cleaning: data validation tools powered by AI, like **Talend** and **OpenRefine**, detect redundant, unconvincing, or incorrect records, thus providing data reliability.
- Real-Time Data Enrichment: Tools such as **ZoomInfo** or **Clearbit** keep your customer profiles constantly up to date, reducing the need for static and old data.
- Standardised Data Governance: By establishing governance frameworks, you can ensure consistency because data is collected, processed, and analysed across multiple teams and systems.

Compliance with Data Privacy Regulations

As data privacy rises in prominence, companies must also adhere to stricter laws and regulations, such as the EU General Data Protection Regulation (GDPR) and the California Consumer Privacy Act (CCPA). There are very real potential legal consequences and reputational damage the company can face if it is not storing the information properly.

- Transparent Data Collection: Organisations should openly report on how they collect and use data about customers, including compliance with global privacy legislation.
- AI-Powered Compliance Monitoring: AI technology in platforms like

OneTrust and **TrustArc** automates compliance verification, verifying that marketing data complies with regulations.
- Anonymisation and Minimisation: AI-supported anonymisation solutions enable companies to store and analyse data in a manner that mitigates data privacy risks but preserves its analytical utility.

Using Zero-Party and First-Party Data

With the increased regulation around privacy, companies are going to have to start leaning more into their zero-party and first-party data collection and less on third-party data. Zero-party data is the information that customers provide with goodwill, like preferences and purchase intentions. First-party data is collected as a result of an individual's interactions with a brand.

- Interactive Assessments/Surveys: Resources like **Typeform** and **Qualtrics** leverage interactive, form filled zero-party energy data and preference-based questions.
- Integration of First-Party Data: AI-based customer data platforms, like **Segment** and **BlueConic**, will collate first-party data from various interaction points that provide actionable customer behaviour.

Using Multi-Touch Attribution for Better Decision-Making

Attribution modelling allows organisations to determine which of their marketing efforts are responsible for the most conversions. The problem with using a traditional last-click attribution model is that they don't fully account for the complexity of today's B2B buying journeys, in which decision-makers are often touched by multiple channels. AI-based attribution tools bring better visibility to marketing influence.

- AI-Driven Attribution Models: **Google Analytics 4** and **Adobe Analytics** utilise machine learning to attribute credit to various marketing touches.
- Cross-Channel Data Integration: With the use of AI-powered solutions such as **Rockerbox**, you can funnel data from digital, offline, and social media to create a pool of attribution sources.
- Predictive Attribution: With the use of AI-powered tools, you can predict what impact your marketing investments will have in the future to drive better budget allocation by businesses.

AI for Anomaly Detection and Campaign Optimisation

Anomaly detection tools powered by AI automatically identify deviations in data trends, which means businesses can respond promptly to potential problems. This functionality is especially helpful in discovering fraud, underperforming campaigns and surprising changes in shoppers' mindsets.

Automated Anomaly Detection: Solutions such as **Anodot** and **DataRobot** leverage automated AI to surface unusual changes in customer activity and sales numbers.

Live Campaign Optimisation: AI-driven marketing platforms like **Persado** and **Pathmatics** are examining live campaign data in order to recommend real-time adjustments for stronger engagement and conversions.

Creating a culture based on data within organisations

Technology is insufficient as an enabler of success in data-driven marketing. It's imperative for companies to develop a culture where data literacy is a priority for marketing, sales and executives.

Data Literacy Training: AI learning platforms like **DataCamp** and **Coursera** for Business offer training courses to enhance marketing teams' data analysis skills.

Cross-Departmental Data Sharing: This promotes the alignment between marketing, sales and customer success teams, guaranteeing that data insights are effectively shared and put to good use.

AI for Decision Support: AI-enabled analytics dashboards such as **Tableau**, **Power BI**, etc., facilitate data democratisation, making insights accessible across teams.

Best practices for data-driven marketing help ensure businesses are strategic, ethical and effective. Through data quality focus, privacy compliance attitude, adoption of the AI-driven attribution model, and building a culture of data literacy, businesses can drive robust marketing while keeping customers' trust.

The Future of AI, Data, and Analytics in B2B Marketing

The application of AI and data analytics in B2B marketing is poised to grow, allowing companies to make decisions that are not only smarter but faster and more strategic. With the development of AI-based solutions, new possibilities will develop around customer understanding, automation of decision-making processes, and the ability to perform personalisation at scale. Yet companies now also need to get ready to grapple with issues around good AI usage, data privacy and keeping up with a fast-moving tech world.

New trends and developments for AI and data-driven marketing

Hyper-Personalisation and AI-enabled Customer Experiences

Personalisation in B2B marketing is moving beyond targeted campaigns into the realm of hyper-personalisation, where rich AI-driven insight equips businesses to customise marketing messages, content and outreach truly at the level of each individual prospect.

AI-powered recommendation engines such as **Persado** and **PathFactory** read natural language and compare the behaviour and engagement history of a customer to deliver hyper-relevant content in real-time.

Dynamic website personalisation platforms, such as **Optimisely AI**, can customise web experiences based on visitor data and intent signals.

Indeed, as AI-powered personalisation continues to develop, brands will be able to provide increasingly frictionless, contextually relevant experiences across every digital touchpoint.

Real-Time AI-Based Decision Making

Real-time analytics and AI-driven decision-making are increasingly core to B2B marketing. With artificial intelligence, teams can view real-time customer interactions and tweak marketing plans on the fly based on predictive information.

Artificial intelligent chatbots and virtual assistants like **Salesloft (Drift)** and **Conversica** interact with leads in real-time, offering tailored responses and nurturing prospects through the sales funnel.

Automated bids in conjunction with a machine-learning-based budget optimisation tool like **Google AI** SmartBidding estimates and automatically adjusts ad spending to maximise conversions based on real-time performance data.

Businesses that adopt AI-powered real-time decision-making will be able to enhance their agility to address shifts both in customer behaviour and market conditions.

B2B Marketing and Conversational AI and Voice Search

As the conversational AI progresses, voice-activated search and AI powered virtual assistants will begin to have a greater impact on your B2B marketing. Content will need to be designed with the aim of being found in AI-led search engines, and conversational AI will need to be integrated into their engagement strategies.

AI-powered voice search optimisation tools such as **Yext** ensure that B2B content ranks effectively for voice queries.

Platforms like **ChatGPT** allow businesses to incorporate AI-based conversation flows into their marketing.

As voice search becomes more mainstream, businesses need to be ready for a transition to AI powered conversational marketing.

The Rise of AI-Created Content and Automated Marketing Creativity

AI is revolutionising content creation in B2B marketing by scaling the way businesses can produce high quality marketing content. AI-generated material is predictably intended to support rather than overtake human creativity, automating the boring bits so that we marketers can get on with our strategies and stories.

- AI content creation software like **Copy.ai** and **Jasper AI** help marketers draft blog posts, emails and ad copy.
- Automated video production platforms, like **Synthesia**, enable companies to make AI-generated video content without physical shooting.

The era of AI content demands that companies strike a balance between automated and human creativity for genuine and engaging content.

Challenges and Considerations in the AI-Driven Future

Responsible AI and Ethical Use of Data

As AI becomes more embedded in marketing, businesses need to uphold responsible AI practices, transparency, and ethical data use. Ensuring that the AI models will be free from bias, comply with privacy laws, and still be checked by humans will be key.

- AI ethics frameworks such as Linux Foundation AI Fairness 360 help businesses identify and address imbalances in AI models.
- Privacy-preserving AI methods such as federated learning give companies the ability to train AI models on decentralised data without revealing private customer information.

Businesses that put ethical AI at the heart of their AI adoption will gain the trust of customers and regulators.

Evolution of Data Privacy Laws

Data privacy regulations will continue to change and influence the way companies source and use customer data. As regulations like GDPR and CCPA expand, brands cannot help but be more privacy-focused in their marketing campaigns.

- AI-based compliance-monitoring platforms such as **TrustArc** and **OneTrust** enable companies to monitor their data practices to ensure compliance with global privacy statutes.
- Tools like **Fathom Analytics** are privacy-focused analytics solutions that give you analytics without storing or collecting personally identifiable information.

Although no less cumbersome in today's regulatory environment, businesses need to transition to a zero-party data strategy, where customers voluntarily provide information in return for meaningful interactions.

Preparing for AI-Driven Marketing Transformation

In order to stay relevant in this AI-powered future, companies should invest in AI talent, flexible analytics infrastructure, and a 'keep learning' mentality. The implementation isn't just about technology; it involves coordination between marketing, sales and IT teams to enable smooth adoption.

- AI learning guides for training programs like **DataCamp** for Business and **Coursera** AI for Marketers help to upskill teams on AI-driven marketing strategies.
- AI integration platforms such as **Zapier AI** are there to help businesses automate their workflows across marketing and sales platforms.

Businesses investing in AI adoption and education will succeed in the data-driven world.

The Next Phase of AI-Driven B2B Marketing

What lies ahead for AI, data and analytics in B2B marketing: greater levels of precision, automation, and personalisation. Businesses that adopt AI-driven decision-making and real-time analytics and apply ethical AI adoption will drive the next wave of marketing transformation.

As AI technologies advance, businesses that embrace this future-ready, data-first focus will have an edge, enabling them to run more successful campaigns, enhance customer experiences, and realise greater marketing ROI. The balance point is that AI-driven automation is the holy grail of where data-driven insight meets a beautiful and more ethical marketing strategy.

The Rise of Attribution in the Age of Dark Social

As buying journeys become less linear, attribution becomes less precise. Classic multi-touch approaches fail to account for the real effect of peer-to-peer messages, podcasts, **Slack** chats and private DMs, all of which now fall into what we now call dark social.

Dark social is the untraceable paths of influence increasingly shaping B2B decisions. One CFO listens to a podcast during their commute. A product lead links to a white paper in a private WhatsApp group. A customer reads a **LinkedIn** comment that doesn't appear in the analytics but changes perception significantly.

It's not just a technical problem. It's philosophical. And yet, most B2B attribution models continue to push against a growing tide of uncertainty. This causes marketers to over-invest in what can be easily measured and under-invest in what really drives belief.

So, how should contemporary marketers respond?

Move away from attributions and toward something a bit softer: Instead of looking at what drives the last click, look at what changes people's perceptions.

Lean on qualitative data: Ask customers questions like, "Where did you first hear about us?" or "Why did you decide to talk to sales?" Open-text responses expose the invisible influence.

Build a model at a macro level: Use cohort analysis, market-mix modelling, and trend overlays rather than individual click paths.

Develop brand awareness with purpose: Embrace that many of the first few impressions (the best ones) can't be measured.

Zero-Click Content: Prioritise for discovery and consumption without a click (e.g., carousel posts, embedded insights, in-feed stories).

Dark social isn't a gap in your funnel. It's proof that marketing is doing what it's supposed to do, creating demand before intent is detectable. Accept it and plan accordingly.

Summary:

Data-driven marketing helps businesses move away from guesswork and towards making strategic decisions that are based on real-time metrics. Using AI and predictive analytics, organisations have the potential to refine their customer targeting their marketing spend and enhance their return on investment.

- Use statistical analysis to develop new marketing strategies and forecast future customer needs.
- Invest in AI-powered analytics tools that provide real-time, actionable insights.
- Create a data-driven culture by building the capabilities of teams to both interpret and apply the data.
- Ensure data privacy and compliance while getting the most out of AI-driven marketing.

Example in Action: Amazon Web Services (AWS) and AI-Driven Customer Insights

Amazon Web Services (AWS) utilises AI-powered analytics to track customer usage patterns, predict demand, and personalise marketing efforts. By integrating predictive insights into its strategy, AWS tailors service offerings based on real-time customer behaviour. This approach has led to increased customer retention and higher upsell success rates, demonstrating how AI-driven data analytics can transform marketing performance.

For more information on AWS's data-driven analytics approach: https://aws.amazon.com/big-data/datalakes-and-analytics/

15. The Infrastructure of Modern B2B Marketing

"If you can't describe what you are doing as a process, you don't know what you're doing."

W. Edwards Deming, the Father of Total Quality Management

Marketing in B2B high-performance organisations is no longer a creative or communication activity; it is a performance engine driven by measurable results, cross-functional alignment, and data-driven decision-making. At the heart of this transformation is the practice of marketing operations and its strategic equivalent, marketing enablement. These "teams" deliver the infrastructure, systems, and governance needed to implement a marketing strategy at scale and in a consistent and accountable manner.

Marketing operations ensure every campaign, from execution to metrics, is efficient, compliant, and integrated across the tech stack. By contrast, marketing enablement focuses on making sure that people, particularly sales and marketing, have the tools, training, and content they need to provide value to the customer at every touchpoint.

They provide the operational scaffolding that underpins modern B2B marketing. Empowered by AI, these functions go beyond the supportive cast; they become the strategic engines of growth and agility.

Defining Marketing Operations in a Modern Context

At its core, the most basic definition is simply the process of planning, governing, and measuring so that marketing teams can run effectively. However, in mature organisations, marketing operations are more than project management. It is the nervous system where intelligence moves, strategy is executed, and performance is measured in near real-time.

The following are key aspects of what is defined as marketing operations:

- Management of marketing technology and automation platforms
- Data Governance, segmentation, and compliance
- Budget tracking, projections, and resource allocation
- Create and update campaign execution workflows
- Attribution modelling and performance measurement
- Innovations in the internal process and quality control

In essence, marketing operations make sure that strategy isn't just ambitious but executable and that execution can be measured and iterated.

Gartner says that as marketing operations maturity grows, so too does business growth performance, with structured operations teams being 2.5 times more likely to surpass revenue targets. https://www.gartner.com/en/marketing/topics/marketing-operations

Marketing Enablement: Driving Alignment and Capability

Where marketing operations focus on systems, marketing enablement focuses on people. It guarantees that internal stakeholders possess the frameworks, advice and content to produce sophisticated and results driven marketing. Enablement in organisations with numerous regions, product lines and customer segments is especially critical, as a scattering of messages and unaligned processes can water down brand effectiveness and hinder cross-functional teamwork.

Key responsibilities include:

- Developing and distribution of playbooks, templates, and style guides
- Sales & Marketing coordination on messaging and campaign strategy
- Role-specific training and onboarding programmes.
- Libraries of content based on personas and buyer journeys
- Development of internal communications and feedback mechanisms

Marketing enablement closes the strategy and execution gap, ensuring teams not only know what to do but also how to actually do it consistently and why it is important.

AI's Effect on Marketing Operations and Enablement

There's a new dimension of strategy in both operations and enablement because of the use of AI. It helps marketing functions to predict proactively rather than react, constantly refine instead of retrospectively, and support decision-making with empirical insight rather than assumptions.

AI in Marketing Operations:

- Campaign outcome prediction: Machine learning models can forecast the likely success of a campaign based on audience, channel and content variables.
- Lead scoring: Predictive algorithms use behavioural, demographic, and firmographic signals to assign scores.
- Performance: AI automatically reallocates budgets, adjusts creative, or changes frequency in flight, based on live inputs.
- Data cleaning and enrichment: Tools like **ZoomInfo** or **Clearbit** enrich the gaps in data and flag whenever data isn't in line.

Platforms such as **Salesforce Einstein** and **Adobe Sensei** embed AI natively, providing real-time recommendations throughout CRM, analytics, and campaign activities.
https://www.salesforce.com/eu/products/einstein/overview/
https://business.adobe.com/products/sensei/adobe-sensei.html

AI in Enablement:

- Conversational intelligence: Companies such as **Gong**, **Salesloft (Drift)**, and **Outreach.io** analyse calls to know what messages, objection handling, and persona resonate best.
- Automated learning: AI provides personalised training paths for marketing staff based on their role, seniority, and performance data.
- Knowledge management: AI-powered search across content libraries, helping ensure teams can quickly find the most relevant collateral.

These are not tools to be used instead of human expertise but instead to complement it. They create consistency, speed up the learning process and maximise cross functionality feedback loops.

CMO Evolving role - Designing Teams for the Age of AI

Today, in the age of AI, B2B marketing is not a department but a system-wide capability. The organisational structure has to be agile.

Emerging Models:

- Pod-Based Squads: Cross-functional teams that work on audience segments, products or life-cycles.
- Growth Guilds: Learning cohorts interested in AI, content, or buyer insight.
- AI Ops Layer: A group that handles Martech infrastructure, governance and experimentation.

Marketing Procurement & Vendor Strategy

The power that marketing holds is increasingly dependent on external partners, creative agencies, artificial intelligence vendors, and so on. But not many B2B teams have a well-honed procurement strategy yet, and that's all right...but that needs to change.

Modern Best Practices:

Capability Mapping: Link vendors to business outcomes, not just capabilities. You can use a visual map to tie together the vendors and your business outcomes, and not just functions.

Scoring AI Tools: Look beyond features and scrutinise vendors on governance,interpretability and ease of adoption.
Agile Contracts: Scopes that are flexible and allow for test-and-learn.
Plan ahead: When purchasing tools, focus on interoperability, data portability, and enabling your team.

How to Scale Your Martech Stack in a Compliant Manner

Martech infrastructure is increasingly responsible for marketing operations. The objective isn't to add a tool or two but to implement them within a network ecosystem that consistently aligns strategic agility, data governance and campaign scalability.

- Some fundamental Martech categories and example suppliers are:
- Automation and CRM: **HubSpot**, **Marketo**, **Pardot**, **Salesforce**

- Analytics and attribution: **Looker Studio, Tableau, Segment**
- Content/asset management: **Bynder, Seismic**
- Integration and automation: **Zapier, Workato, Tray. io**
- Planning and collaboration: **Adobe Workfront, Monday. com**

All of these need to work together under a common data model, fit together seamlessly and comply with governance. The marketing operations function is the gatekeeper – striking this balance between innovation and oversight.

Privacy regulations like GDPR, PECR, and CCPA make data governance an essential aspect of our business. AI tools need to be trained, deployed and monitored with an eye on privacy laws and ethical guidelines. Frequently, this responsibility rests within operations, working alongside legal and data teams.

From Marketing Ops to Revenue Operations (RevOps)

We're now seeing the next evolution of marketing operations, its convergence with sales and customer success functions into a unified Revenue Operations (RevOps) model. This integrated approach eradicates institutional silos, harmonises data and focuses teams on a common collection of measurements.

In a RevOps model, marketing operations broadens its focus to growth:

- Common pipeline and revenue metrics
- Cross-functional attribution frameworks
- Unified tech stacks and automated workflows
- Centralised business intelligence dashboards
- Governance throughout the customer life cycle

It does need a big mental shift, but the performance gains are thereby eliminating the silos between departments.

Avoiding Common Pitfalls

When marketing operations and enablement are fully mature, they usually face the following challenges:

- Tool sprawl: too many systems with redundant capabilities lead to inefficiencies.

- Data debt: Bad data hygiene erodes campaign accuracy and accountability
- Low usage: Teams don't use tools because they are undertrained or the experience isn't there.
- Siloed planning: Marketing, sales, operations on different timescales
- Measurement gaps: KPIs that aren't tied to business results

These can all be resolved through proper governance models, internal comms strategies, and a cross-functional Centre of Excellence that is responsible for both implementation and innovation.

Case Example: The Evolution of Global Marketing Ops at Cisco

Cisco Systems initiated a large-scale makeover of its worldwide marketing to accommodate more agile, data-driven plans. The company spent about $50 million in marketing software, rolling out around 40 new applications to assemble a complex Martech stack. This move was part of a broader strategy to transition Cisco to a software and services-oriented organisation - mirroring the fact that business purchasing habits have started to emulate those of consumers.

With tools powered by AI embedded across the stack, Cisco can now track worldwide campaign performance and shift tactics locally while aligning sales and marketing with unprecedented precision. This transformation has resulted in higher lead conversion rates, faster delivery times for campaigns and a more consistent global brand. https://web-assets.domo.com/blog/wp-content/uploads/2018/10/Cisco-Digital-Marketing-Transformation-Case-Study-1.pdf

Additionally, Cisco partnered with Adobe to modernise its digital experiences, including the migration of Cisco.com from an on-premises Adobe Experience Manager environment to Adobe Experience Manager as a Cloud Service. https://business.adobe.com/uk/customer-success-stories/cisco-case-study.html

These modernisation efforts have allowed Cisco to more precisely orchestrate its customer journeys, which in turn enable marketing and sales to increase mindshare and drive more sales and, ultimately, more valuable customer outcomes.

Summary:

Marketing operations and enablement are no longer "nice to haves" in today's B2B marketing world. They dictate how well a plan is executed, how nimbly teams can adjust, and how objectively performance can be assessed. In an AI-enabled world, these processes need to be proactive, strategic and closely aligned with revenue results. B2B companies that promote the operational maturity of their operations today will have the speed, transparency and agility to excel tomorrow.

16. Going Global with AI: Strategy, Scale and Sensitivity

"Success is best when it's shared."

Howard Schultz, CEO of Starbucks

B2B marketing on a global scale has changed, and companies need to coordinate cultural nuances, decentralised regulations, and shifting regional market dynamics while staying true to their brand. Traditional global marketing strategies were also typically based on manual localisation processes. They were developed with broad segmentation models, which were rarely flexible enough to cope with fluctuating international market trends. However, AI is changing these dynamics with the arrival of automated and real-time analytics and predictive modelling, allowing businesses to make better decisions using hard data.

The dramatic improvements that AI has been able to bring to marketing are already evident according to a report by **McKinsey & Company**, which found that AI-powered marketing applications are reaping the rewards as businesses that have begun using AI in their marketing at a global level are seeing a two times increase in the rate of engagement, leading to a 3–7 percent increase in conversion rates and 9–19 percent stronger customer satisfaction rates. AI allows businesses to constantly adapt marketing strategies with real-time market intelligence to ensure messaging, branding, and customer engagement efforts effectively target regional audiences.
https://www.mckinsey.com/capabilities/growth-marketing-and-sales/our-insights/ai-powered-marketing-and-sales-reach-new-heights-with-generative-ai

The Shift from Traditional to AI-Powered Global Marketing

Traditionally, worldwide marketing plans were based on large-scale regional segmentation, geared around past data, established localisation approaches and quarterly performance auditing. Marketing organisations frequently crafted campaigns for specific markets and incorporated information mined from demographic studies, focus groups, and historical response data. Although these methods offered important guidance, they were mostly reactive and took a long time to respond to evolving consumer behaviours in real-time.

AI is revolutionising this approach by providing real-time market intelligence, automating localisation, and consistently optimising campaign delivery. AI-enabled applications can interpret consumer sentiment worldwide, monitor new shifts in engagement, and automatically tailor content based on what's 'hot' in a given geography. With AI-powered insights, businesses can now predict when customer behaviour changes, align messaging and develop campaigns that adapt on the fly based on actual responses.

AI's Role in Adapting to Regional Market Conditions

AI greatly improves global marketing capabilities through the ability to track real-time regional trends, cultural sentiment and customer tastes. Everything from sentiment, from social media comments to customer feedback to market performance data, is collected and analysed by machine learning algorithms that offer marketing teams usable insights. Advanced machine learning-based predictive analytics helps companies predict demand in various regions so marketing tones can resonate with changing consumer sentiments.

Research from **McKinsey** Digital indicates AI will transform a range of business functions, from marketing to sales to customer engagement, by boosting productivity and providing highly personalised experiences at scale. Businesses using AI-based market adoption tools are in a better place to fine-tune pricing strategies, product placement strategies and promotional efforts based on regional data.

https://www.mckinsey.com/capabilities/mckinsey-digital/our-insights/the-economic-potential-of-generative-ai-the-next-productivity-frontier

With AI's ongoing evolution, companies that incorporate AI-enhanced insights into their global marketing strategies are going to be miles ahead. Automating localisation, monitoring live engagement patterns and predicting the shift in demand in a region allows marketers to take action in a more relevant and timely way than ever before.

Localisation and Market Tuning using AI

Popping up in international markets is more than just translating content. Effective localisation not only helps to get marketing messages right for local audiences but also considers cultural idiosyncrasies, local idioms, and market-fed trends. Conventional localisation used to be mostly done manually by translation, regional marketing teams, and extensive market analyses. Although these approaches yield meaningful results, they are often slow and inflexible in coping with dynamic customer expectations. Today, AI is redefining localisation, allowing you to scale your global marketing campaigns and keep your messages culturally relevant and 'on brand'.

According to research by CSA, 76% of international online shoppers want to buy products in their own language or with information provided in their native tongue, despite the quality of the translation. This only serves to underscore the criticality of effective localisation, not just mere translation, whereby tone, imagery and regionally specific messaging are taken into account. At scale, AI-powered localisation platforms enable businesses to do this by automating content localisation while also maintaining the brand message. http://csa-research.com/

AI-Powered Localisation Platforms and Their Impact

Artificial intelligence localisation tools have transformed the way organisations deal with content in multiple markets. These solutions leverage machine learning and natural language processing to drive automated translations, polish messaging, and guarantee the context is always relevant.

DeepL is an AI-based translation service that does more than simply translate word for word; it is best for adapting marketing content.

Smartling integrates AI with human translation workflows, ensuring that content remains accurate while being culturally relevant to each target market.

Unbabel pairs the speed of machine translation with human editing so businesses can create high-quality content in multiple languages.

With localisation platforms like the above, powered by AI, your company can open itself to international business opportunities and maintain messaging that is not only effective in every territory but also engaging. With AI, companies can track linguistic shifts, regional sentiment, and audience preference in real-time and then adjust content on the fly according to the way those markets respond.

Cultural Adaptation Through AI-Driven Insights

However, effective localisation goes beyond language and, therefore, involves cultural adaption to ensure that marketing materials match local practices, values, and expectations. AI-infused sentiment analysis tools enable companies to gauge how local audiences will feel about particular messages so they can tune campaigns before going live.

- **Brandwatch** monitors social media communications, forums, and other online conversations by region, making it possible for boutiques to gauge how people in different cultures feel and adjust their messaging.
- **Sprinklr** provides AI-driven insights into customer behaviour and cultural trends, enabling companies to tailor their marketing strategies to local preferences.
- **Lexalytics** uses natural language processing to assess customer sentiment, identifying potential cultural misalignments in marketing content.

Visual localisation is also of paramount importance for AI. Cultural preferences vary in terms of colour symbolism, imagery, and design aesthetics. AI-powered content adaptation platforms, like **Adobe Sensei**, analyse regional engagement trends to suggest visual elements that are most likely to resonate with certain audience segments.

Dynamic Global Campaigns Making the World Local

Real-time content adaptation by AI is one of the most important advantages it brings to localisation. Every campaign can be unique according to the context, using AI-driven content placement to target specific industry needs with locally relevant data and data that tracks engagement to know what customers care about.

- Optimisely AI dynamically customises website content and ad creatives according to user location, device, and behaviour, providing a unique experience in every market.
- **PathFactory** provides AI-driven content recommendations tailored to regions to maximise engagement through personalised user journeys.
- **Persado** brings machine learning to bear on understanding the emotions and language cues that best perform in a variety of markets, enabling companies to craft powerful and locally appropriate marketing messages.

AI is changing how companies think about localisation, enabling them to scale global marketing while still being true to local culture. With AI-driven translation tools, sentiment analysis platforms and real-time content adaptation as part of the content marketing mix, companies can adapt conversations and stories in real-time to reach new and niche audiences. This ensures marketing in any culture isn't lost in translation while also enabling businesses to expand internationally.

Compliance, Privacy and AI-Driven Solutions

When working in and crossing international markets, large geographic areas of potential datasets are subject to vastly differing regulations on data privacy, consent and marketing compliance. With the European Union's General Data Protection Regulation (GDPR), the California Consumer Privacy Act (CCPA) in the US, and Canada's Personal Information Protection and Electronic Documents Act (PIPEDA), B2B marketers with a global footprint must make sure that their practices comply with the legal standards of each jurisdiction in which they operate. Not complying can result in hefty fines and harm to the brand name.

Increasingly, artificial intelligence is being leveraged to assist businesses with this complicated regulatory environment. With automation of compliance checks and managing data flows, AI helps firms maintain compliance and scale marketing operations across the globe using consent tracking across regional boundaries.

The Challenge of Global Data Privacy Compliance

There are many different data protection laws for each nation or region. Though GDPR has established a global standard in privacy regulation, other countries have since adopted their own versions of GDPR-like legislation. This has created a patchwork of regulations, which may be unwieldy to manage without the right tools.

Marketers should consider needs in terms of:

- How we collect and store data
- Opt-in choices and consent tools
- Cross-border data transfers
- Customer's rights to access, modify or delete personal data

Conventional compliance activities are dominated by legal oversight and manual resource-intensive and error-prone processes and procedures. AI allows for a transition from reactive compliance management to proactive risk mitigation.

AI-Driven Compliance & Consent Management Solutions

Some platforms have begun to leverage AI to automate compliance monitoring and data subject rights, as well as meet global privacy values in real-time.

- **OneTrust** provides all-in-one cookie consent, data mapping, and privacy impact assessments with an AI engine to keep abreast of regulatory updates within regions.
- **TrustArc** applies machine learning to simplify global privacy compliance, automate risk assessments and provide real-time risk reporting for multinational marketing teams.
- **BigID** uses AI for data discovery and classification so that companies can know and secure what personal data they have across their systems.

They do more than comply with shifting requirements; these platforms free up time (by lessening manual requirements for compliance) so marketing teams can concentrate on strategy and execution.

Cross-Border Data Transfers and AI-Based Safeguards

Cross-border data transfer is a paramount concern of the global marketer. Laws frequently prescribe a strict level of adequacy regarding data leaving a specific country or that it should be safeguarded by mechanisms such as Standard Contractual Clauses (SCCs).

One such solution is to use AI to automate the detection of the risks of transferring the data and suggest safeguards. This includes:

- Automatic data lineage monitoring to follow the origin and the flow of data
- AI-driven contract analysis for missing regulatory clauses
- Data exposure and breach risk alerts in near real-time regardless of location

In more sophisticated uses, blockchain is seen as a way to keep transparent and tamper-proof trails of data, which can show businesses that data has been managed according to local law.

Striking the Right Balance between Personalisation and Data Ethics

With so much personalisation and targeting via AI, it's important for marketers to remain compliant and respectful in utilising AI's capabilities for data use. AI models need to be responsibly trained to avoid biased outputs, and data treatment has to be done in a manner that retains trust and transparency.

Businesses that employ at least one AI-based personalisation tool also need to:

- Define internal policies on the use of responsible AI
- Conduct regular audits of algorithms for fairness and bias
- Offer users the ability to curate the degree to which their experience is data-driven personalisation

Open-source toolkits like the Linux Foundation AI Fairness 360 sit beside the computer infrastructure on which machine learning systems are built and help companies be confident that their AI-based marketing campaigns are operating in an ethical way. https://ai-fairness-360.org/

AI and Regulation Change Management

One of the most dynamic aspects of compliance is keeping pace with changing regulations. AI comes to the rescue, tracking changes in privacy laws across the globe and automatically updating internal policies. Some platforms also offer real-time legal intelligence feeds, similar to news updates, which advise marketing and legal on any changes and regional enforcement activity on the horizon. Privacy-aware analytics tools like **Piwik PRO** or **Plausible** are now available.

This adaptability enables businesses to:

- Predict regulation changes and be ahead of enforcement efforts
- Don't let legal review hold back campaigning start dates
- Log your regulatory due diligence and historical process trails

By way of summary, global B2B marketing is today having to work within an evolving and complex legal landscape. AI is the toolkit to automate compliance, police ethical data use, and enforce the same standards across billions of international data flows. Adopting AI-powered technology solutions can help organisations reduce their likelihood of liability and respond more quickly to evolving privacy trends when delivering personalised, compliant experiences everywhere they do business.

AI-Augmented Strategies for International Campaign Management

Running international campaigns is about way more than scaling a message across multiple markets. Achieving this demands that global and regional teams work tightly together, that brand execution is seamless, and that the brand can flex to local tastes, behaviours, and conditions. Multinational campaigns in the past used to be run on manual coordination, fragmented reporting and reactive strategy changes. They had been slow and cumbersome and did not scale, creating inefficiency and lost opportunities.

Artificial intelligence is revolutionising international campaign management by providing better visibility worldwide and supporting predictive planning and campaign automation. By harnessing the power of AI, companies can achieve a harmony of global strategic vision and regional entrepreneurial freedom, creating campaigns that are more integrated, effective, and locally resonant.

Coordinating Worldwide Strategy and Local Follow Through

One of the biggest problems with international marketing is actually keeping global and local staff in sync. So, while global marketing gives you the brand strategy, local teams will execute the campaigns according to your local market and preference. AI makes this partnership more dynamic and data-driven.

Collaboration platforms with AI like **Asana** Intelligence and **Monday. Com,** AI can now help plan and track campaign performance and guarantee regional teams adhere to brand guidelines but still be nimble enough to customise content to local market conditions. These tools make it possible to see in real-time the timelines, assets, and budgets for the campaign in different geographies to ensure consistency without sacrificing responsiveness locally.

AI-Powered Campaign Optimisation at Scale and Across Markets

AI capabilities can be used to track and analyse the performance of campaigns by market, highlighting poor performers for intervention and recommending optimisations based on historical customer response patterns and customer engagement and behavioural signals. This gives marketers the ability to act based on evidence, iterate on messaging, and transfer spend to where it has more of an impact.

Adobe Experience Platform leverages AI to deliver unified customer profiles and cross-channel analytics so marketers can customise messages and offers by region using real-time performance data. https://business.adobe.com/uk/products/experience-platform/adobe-experience-platform.html

With **Salesforce** Marketing Cloud Intelligence, AI is used to provide insights into campaign stats on platforms and across regions, enabling marketers to understand which strategies are most effective by location. https://www.salesforce.com/marketing/analytics/introducing-marketing-intelligence/

Cross-regional campaign optimisation, suggestions for reallocating budgets, and automated reporting using AI are capabilities under the **Google** Marketing Platform. https://marketingplatform.google.com/intl/en_uk/about/

Businesses can use predictive models to predict the results of a campaign before it has launched, minimising risk and ensuring messaging is tailored regionally.

Real-Time Sentiment Monitoring and Culture Calibration

AI is also critical to understanding how global campaigns are perceived and responded to in various regions. Soft metrics monitoring, tracked in real-time with software like **Brandwatch** or **Talkwalker**, are designed to give marketers an early read on whether a campaign is striking the right note or needs to be recalibrated with regard to cultural dissonance or tone.

How brands use sentiment insights:

- Tailor content and visual elements to local emotional sensibilities and cultural expressions
- Steer clear of mistakes that touch on local sensitivities or history
- Localise the global story and find local storytelling opportunities

This kind of proactive monitoring keeps reputations on track and campaigns feeling local - even when they're running globally.

Example: Cisco's AI-powered Global Campaign Management Program
Cisco provides a strong example of how a multinational enterprise can use

artificial intelligence to harmonise global and regional campaign execution. According to **Cisco**'s own case study, the company built and deployed an AI-powered framework that supports over 25 use cases, including real-time analytics, automated content delivery, and enhanced campaign oversight. This infrastructure allowed **Cisco** to accelerate campaign timelines by 80 per cent and respond more effectively to regional market demands. https://www.cisco.com/site/us/en/solutions/cisco-on-cisco/ai-ready-infrastructure.html

Incorporating an AI-powered marketing solution into its global marketing operations, **Cisco** has improved its ability to monitor how content is performing, change the direction of campaigns in real-time, and deliver regional teams with new tools and insights so they can run locally applicable campaigns.

Growing In Size Without Losing Relevance

Generalisation is also a rampant concern in international marketing – far too often, campaigns don't scale efficiently, and communication becomes confused and diluted. AI enables marketers to scale with intelligence and, in doing so, preserves the precision of local insights while harnessing shared data sets and predictive models to inform content, offers and channel strategies.

- AI is available to recommend the most advantageous channels to use in different places based on local behaviour patterns.
- It can even customise email or page content by language, industry, or regional business culture.
- Machine learning models help identify the right times and frequency for messages in various timeframes in time zones.

This degree of optimisation would be too much to handle via manual rules on as many markets as possible.

Striking The Right Balance Between Localisation And Brand Consistency.

Balancing Localisation and Brand Consistency

One of the biggest global marketing challenges is keeping your corporate identity intact while messaging to different regional markets. Too much centralised control from marketing could produce irrelevant or tone-deaf messages at a local level, and overly localised content may water down the brand, its messaging and its impact. Artificial intelligence is now being utilised to get it right, providing adaptive content creation, dynamic messaging oversight and real-time brand guideline monitoring across regions.

Consistent branding builds trust, awareness and credibility. However, global readership varies in language, culture, values, and business customs. The key to international marketing success is the capacity to harmonise these contrasts without sacrificing brand consistency. Today, AI is enabling companies to monitor, guide, and align messaging in every market.

Artificial Intelligence Brand Governance Tools

Brand governance AI solutions guarantee every piece of marketing content aligns with your organisation's brand voice, tone, and identity, irrespective of the channel or geography. This ranges from ads and landing pages to sales enablement content and social media content.

- **Acrolinx**'s natural language processing platform checks and ranks content against brand standards, allowing companies to scale the consistent delivery of messaging no matter where or how it's being deployed.
- **Grammarly** Business provides writing assistants and real-time suggestions to support content creators in maintaining a cohesive tone, clear messaging and consistent branded language company-wide, across all teams, and across international contributors.
- **Contentful** integrates with AI systems, enabling version control and content templates across multilingual deploys and enabling global teams to create localised assets with the global framework.

These solutions really shine when they sit within a global marketing workflow, offering regional teams a set of rails to stay on while having flexibility in the creation of content.

Dynamic Branding and Customised Messaging

Dynamic branding can be achieved through AI, allowing campaigns as diverse as headlines, images, and calls to action to automatically adjust according to locally relevant interaction, preference, or cultural trigger while ensuring brand consistency.

- **Persado** uses AI to generate emotional marketing language that reflects your brand voice and adjusts for local conditions.
- **Phrasee** provides AI-optimised email subject lines and advertising copy, in real-time and at scale, to enterprise brands across the globe, saving time and making marketing language more effective, predictable, and accurate in any language.

Dynamic branding ensures the messaging of the business is not adjusted from its core messaging and visual identity yet offers any nuanced content or meaning that may matter to the consumer.

Implementing the Visual Identity at the Local Level

Brand consistency is as much about what you see as what you say. Colours, iconography, layout conventions and even white space emphases may differ from region to region. So, tools are available now to optimise the creative with brand style guidelines in place.

- **Canva** for Teams enables global marketing teams to design locally with locked brand templates that adhere to brand guidelines.
- **Adobe Firefly** makes it possible to create AI-generated visual content locally within brand design rules, bringing commercially safe localisation.

With the use of AI-powered design systems, companies are able to decentralise creative work yet maintain visual consistency across all customer touchpoints.

Global Brand Audits Using AI

Sustaining a brand in different countries and with other cultures also demands constant control of execution. Artificial intelligence enables this process by reviewing website content, sales materials, social posts, and digital assets for adherence to brand standards.

- Brand audit tools that leverage AI can identify content inconsistencies suggest enhancements and notify regional teams if alignment falls below the acceptable threshold.
- Machine learning models can find patterns of misalignment across areas and drive training and process improvements for local teams.

This is a proactive way to ensure quality at scale for marketing leaders without getting stuck in the weeds on every output.

With AI, businesses are now tackling the difficult balance between localisation and brand consistency. AI-driven governance, dynamic content adaptation and continuous brand monitoring mean businesses can empower regional teams to innovate and act locally, all while protecting global brand cohesion.

The Future of AI and Global B2B Marketing

As AI evolves, so will its place in global B2B marketing – meaning more precision, automation, and strategic "smart" decisions. AI is already changing the way businesses handle localisation, compliance, campaign running and brand consistency. The next horizon is for AI to become an integrated decision-making partner, anticipating market shifts, identifying emerging trends and continuously optimising strategies across every touchpoint.

The enlightened B2B marketer is planning for the future where AI isn't just the tactical execution support of their wildest dreams but is interactively guiding top-line business strategy in real-time.

Global Marketing Automation Trends in the Making

AI platforms are becoming more advanced, and they are graduating from rule-based automation to self-learning systems that optimise campaigns on the basis of ongoing feedback.

Key developments include:

- Generative AI for scalable content creation: AI is going to continue to play a larger and larger role in generating region-specific marketing copy and even imagery and video content, meaning we'll start seeing assets created at scale at significantly faster paces and more personalised for regional tastes. Some tools like **Jasper AI** and **Synthesia** already make this possible.

- Real-time localisation at scale with multilingual AI models: Brands can localise their tone, messaging and message delivery in seconds across cultural contexts and large language models (LLMs) based on user behaviour, sentiment data and campaign engagement.
- Hyper-personalised journeys: More and more AI will back journey orchestration engines that create fully personalised experiences across websites, emails and ads, tailored by language, market maturity, industry and even cultural buying behaviours.

Putting Voice, Visual and Conversational AI Together

The future of marketing is not just in automating but in natural, human-like interaction. With conversational AI, companies are now able to interact with customers in real-time, using the customers' language and tone at every stage of the buyer's journey.

- AI-driven multilingual chatbots from the likes of **Intercom** or **Salesloft (Drift)** provide 24/7 engagement across different markets, instantly changing language preferences and adjusting responses based on local context.
- Voice search and smart assistant optimisation will gain traction, especially in emerging markets where mobile-first is the norm. Companies need to start optimising content for voice interfaces and AI technology like **Google** Assistant and **Amazon** Alexa.
- Visual AI and image recognition will empower marketers to dynamically adapt creatives to target markets' cultural norms and visual tastes, increasing relevance while maintaining brand consistency.

The Emergence of Predictive Global Strategy

With the continued evolution of predictive analytics tools, companies will transition from reactive to proactive marketing. AI will even be used to determine under-served markets, forecast growth trends in specific regions, and suggest budget reallocation across regions.

Already, companies like **6sense** and **Demandbase** are using AI to predict where buyer intent is spiking in order to target spending with greater precision.

In the near future, marketing leaders will have AI dashboards that they look at not to see how well their marketing is doing but to perform real-time scenario planning, regional forecasting and a dynamically shifting portfolio, giving life to a global marketing operation that is really agile.

Ethical Considerations and Deployment of Responsible AI

As AI systems make more decisions, businesses need to enable transparency, fairness, and accountability. Technical innovation will drive global marketing in the future, but so, too, will ethical leadership.

Key focus areas will include:

- Elimination of bias in AI models to guarantee no cultural stereotyping or discriminatory assumptions in decisions around content and targeting.
- Data use transparency gives customers in different regions transparency and control over how their data is used for AI-powered marketing.
- Governance models that harmonise AI across an international set of team members and guarantee consistency, compliance and ethical oversight.

The emerging field of AI ethics in marketing is being facilitated by organisations like The Alan Turing Institute and tools like Linux Foundation AI Fairness 360, which offers companies a way to do so in a way that's more suitable to scale.

Getting Ready for the AI-Native Marketing Future

To stay ahead of the competition in an AI-led global marketplace, B2B marketers must start to invest now in:

- AI education and training on marketing, sales and analytics staff
- Infrastructure and data governance for AI
- A series of pilot programmes to see how AI-powered personalisation, localisation, and automation tools can work
- Cross-functional discussions from Marketing, IT, Legal, and regional perspectives to be able to form an integrated AI strategy

Businesses that see AI as a marketing discipline, not just another tool, will be the torch-bearer of next generation global B2B innovation. There is no question that the future of global B2B marketing is reliant on successfully integrating artificial intelligence into every business function and market. From adaptive messaging and content creation based on user behaviour to predictive analytics and live customer engagement, AI is changing the realm of what's possible globally.

Those who know that AI is here to stay will not just keep up but will lead. The winners in global marketing will be those who tap AI not to replace human creativity but to augment it, producing rich, localised experiences that resonate with buyers at every corner of the globe.

Summary:
Successful global B2B marketing needs to find a balance between branding needs and localisation. Companies with regional knowledge that effectively reposition content with cultural relevance and take advantage of AI-generated in-market data will have the upper hand in diverse markets.

- Adapt campaigns to respect cultural, linguistic and regional preferences while maintaining global brand consistency.
- Create high collaboration with national and regional marketing teams to ensure successful execution.
- Unlock efficiencies with AI-powered automation, sentiment analysis and instantaneous data insights over thousands of global campaigns while enhancing local engagement.

Example: Caterpillar's AI-Powered Localisation Strategy
Caterpillar's global marketing approach has been skilfully localised to be sensitive to individual local markets yet remain forceful and unified. The company uses AI-powered data analysis to match messaging to local cultures, industries, and customer requirements. With the power of local intelligence and predictive analytics, Caterpillar sets up campaigns that are relevant and engaging from market to market, enhancing its claim as the world leader in industrial equipment marketing.
https://www.caterpillar.com/en/company/strategy-purpose/strategy.html

17. Purpose-Led Marketing: Ethics, Impact and Trust

"It takes 20 years to build a reputation and five minutes to ruin it. If you think about that, you'll do things differently."

Warren Buffett

Sustainable B2B marketing is no longer a box-ticking exercise or something notable on a company's 'About Us' page. It's becoming more and more strategic with changing buyer expectations, regulatory scrutiny and investor mandates. Procurement teams are being pushed to align themselves with environmental, social and governance (ESG) frameworks, and stakeholders increasingly see sustainability performance as indicative of longer-term organisational resilience.

One of the things which B2B marketers need to do in this instance is to widen their remit. You can't get away with just selling features and benefits anymore. The new B2B brand needs to articulate its value, demonstrate commitment, and elevate sustainability to a real point of differentiation in a crowded market of purchasing options.

Artificial intelligence only accelerates the need for that change. However, as AI unlocks efficiency and insight, it also raises new questions about energy use, algorithmic discrimination, and data ethics. So, it isn't just about marketers using AI in a responsible manner; they also need to be able to prove how they are following technology decisions that align with broader ESG principles.

The B2B Business Case for Sustainability

For years, purpose-led branding has been a mainstay of consumer marketing, although B2B businesses are now getting in on the act. Seventy-nine percent of B2B buyers are likely to switch to more environmentally sustainable suppliers, according to a recent **Capgemini** study. https://www.capgemini.com/us-en/insights/research-library/the-future-of-b2b/

It's this change that has moved sustainability from a peripheral consideration for purchasing to a central pillar of commercial strategy. ESG considerations are now commonly included in tender requirements, partner evaluations and even investment pitches. For marketers, this is not just a matter of messaging; it is a way through which brand, value proposition and differentiation must be rethought.

For organisations that integrate ESG into their operations and can authentically and measurably communicate this, the potential to establish trust, preference and long-term loyalty is significant.

How to Include ESG When Defining the Brand

The most successful business-to-business companies no longer consider ESG as a compliance box to check. They are embedding sustainability within their brand architecture rather than adjacent to it, allowing it to play as much a role in differentiation as in driving commercial performance.

Take **Siemens**, for example. Its brand positioning brings together decarbonisation, circular economy initiatives, and ethical supply chain practices, supporting both government and enterprise customers who are increasingly demanding evidence of ESG compatibility. https://www.siemens.com/global/en/company/sustainability.html

Big consumer brand **Unilever** has shown that purpose-led messaging drives real commercial value not only in the consumer world but also in B2B sectors, including food service, logistics and wholesale distribution. Its commitment to sustainably sourced emissions reductions and reuse of plastic is not merely documented – it is quantified, third-party verified, and built into sales enablement content. https://www.unilever.com/planet-and-society/

The implication for B2B marketers is clear. For ESG claims to be effective, they need to be operationally based, well communicated and materially relevant to the buyer's context. The risk of greenwashing, or even its perception, can be ruinous in procurement-led settings used for due diligence.

Earning Trust with Certification and Transparency

As sustainability claims proliferate, trust becomes the currency. Buyers, regulators, and investors increasingly demand third-party validation, not just internal reporting.

Marketers must make sure their claims are backed up by recognised certifications and standards like:

- ISO 14001 – for environmental maintenance
- B Corp status on social and environmental performance
- Science Based Targets initiative (SBTi) for emissions
- CDP (formerly known as the Carbon Disclosure Project) scores on carbon disclosure and climate risk
- GRI (Global Reporting Initiative) for ESG reporting requirements.

Transparency, however, is just as important. B2B customers increasingly require ESG reporting in the tender process, and they use platforms such as **EcoVadis** or **Sedex** to benchmark suppliers. Marketing will need to be able to engage with technical stakeholders who will examine such claims through a compliance and risk filter, not just brand.

AI in the context of Responsible Practices

AI can assist in meeting sustainability objectives, but only if implemented thoughtfully and with oversight. It also poses new risks that marketers will need to navigate carefully.

Where AI adds value:

- Operational effectiveness: Streamlining supply chain logistics to lower emissions

- Intelligent resource prediction: Forecasting energy consumption or travel demand
- Data integration: ESG metrics for integrated sustainability dashboard
- Modelling impact: Scenarios for investing in sustainability

However, the carbon impact of AI itself, especially Large Language Models (LLMs) and data centre usage means companies should monitor and report the energy consumption of AI-powered campaigns and operations.

Tools like the OECD AI Principles help identify and address bias. https://oecd.ai/en/ai-principles

The aim is not to step back from AI but to ensure that its use is consistent with the organisation's values and that its customers can trust how their data is being used, interpreted, and acted on. See the OECD first round of reports submitted through the G7 Hiroshima AI Process Reporting Framework. https://oecd.ai/en/wonk/initial-submissions-g7-hiroshima-ai-process-reporting-framework

Regional and Cultural Factors

There is geographical variation in sustainability preferences.

For instance:

- Climate risk disclosure and supply chain transparency are progressively being regulated in the EU
- North America – focus might veer more towards innovation and ethical governance
- In the APAC region, social impact and employee well-being emerge as "rising themes", particularly in the manufacturing environment.

Marketers will need to make sure that global messages allow for regional variations. This may also involve localising ESG content, adapting language or showcasing regionally specific initiatives that make cultural and economic sense.

Strong collaboration with in-market teams is key to strike the right balance between a unified global brand and credible local appeal.

Internal Alignment: Marketing and the ESG Office

In most organisations, marketers work apart from an ESG or sustainability team. This results in missed opportunities for integration, messaging gaps and reputational risk.

Marketing leaders should:

- Establish communication with sustainability and legal teams at a regular frequency
- Agree on approved language, metrics, and certifications
- Partner on reporting cycles to give marketing visibility upfront
- Be involved in ESG governance committees as applicable

It's not just message alignment but a partnership to make sure that ESG performance is seen, understood, and activated through marketing strategy. Sustainability in B2B marketing isn't taking on a cause; it ties into the direction of commerce's future. Generic claims or empty promises don't cut it with consumers. They want evidence, accountability and a long-term vision.

For B2B marketers, this is a challenge and an opportunity at once. Those who are the most aggressive in embracing ESG as a means of differentiation, using AI responsibly and communicating transparently and consistently with buyers will not only build and preserve buyer trust but also put their brands in good stead in a world that increasingly values responsibility even as it seeks innovation.

Summary:

Sustainability has shifted from compliance to competitive advantage in B2B, driven by ESG mandates, buyer expectations and procurement scrutiny.

- B2B marketers must elevate ESG from messaging to strategic brand positioning, proving value through operations, not just intentions.
- AI accelerates both opportunity and risk, demanding transparency in usage, energy footprint, and ethical alignment with ESG principles.
- Third-party certifications (e.g., ISO 14001, B Corp, SBTi) are now vital tools in earning trust and procurement-level credibility.
- Global variations matter: ESG themes and regulatory expectations differ by region, so messaging must adapt without diluting the core brand purpose.
- Close collaboration with ESG teams is essential to align metrics, messaging and risk visibility, preventing gaps and reinforcing authenticity.
- Marketing's role is evolving from storytellers to stewards of trust, guiding how sustainability is seen, measured and believed across buyer journeys.

Example in action: DataToBiz Enhances ESG Performance for a Multinational Consumer Goods Company

In 2024, **DataToBiz** collaborated with a multinational consumer goods company headquartered in New York City to transform its Environmental, Social, and Governance (ESG) performance. The company faced challenges with fragmented ESG data across its global offices, leading to inconsistent reporting and difficulty in tracking sustainability metrics.

This transformation not only improved the company's ESG ratings but also bolstered stakeholder trust and positioned the company as a leader in sustainable practices within the consumer goods industry. https://medium.com/%40kavika.roy/datatobiz-case-study-transforming-the-esg-performance-of-a-multinational-consumer-goods-company-in-96f801162c21

18. Proving Impact: Metrics that Matter

"If you can't measure it, you can't improve it."

Peter Drucker, The Pioneer of Modern Management

Peter Drucker's philosophy is especially important in the B2B marketing side of sales, where decision-making time and customer relationship building can take longer. Tracking marketing efforts can also be hard to measure without the proper tool set. Knowing how effective your marketing campaigns are is no longer a luxury but a necessity to ensure the best return on investment and to achieve profitable business growth.

B2B marketing is evolving, and digital transformation is driving that change alongside artificial intelligence, revolutionising how businesses monitor and improve performance. Conventional measurement practices, like monitoring typical lead generation and revenue attribution, are being superseded by AI-optimised analytics that offers real-time insights, predictive modelling, and automation. Using AI to measure performance means that companies can stop looking in the rear-view mirror and instead move forward with more informed decision-making and campaign optimisation.

According to a study released by **McKinsey & Co**, firms that use AI to enhance marketing analytics are 15 to 20% more efficient and gain better precision in performance benchmarking. These innovations enable companies to converge marketing goals with revenue growth, allocate resources more efficiently, and personalise customer engagement. https://www.mckinsey.com/capabilities/mckinsey-digital/our-insights/the-economic-potential-of-generative-ai-the-next-productivity-frontier.

This chapter explores how B2B marketers can measure and optimise performance efficiently through AI-powered analytics, sophisticated attribution models, and continuous iterative improvement approaches. It also investigates the extent to which businesses can incorporate agile marketing practices to better drive decisions and maintain success.

Performance Metrics and AI-Powered Analytics

Historically, marketers paid attention to lead generation, conversion rates and revenue attribution. Though they're still important, AI is changing how businesses analyse and leverage marketing data, providing real-time insights, making predictions and telling you to change a strategy before you've failed, and executing decisions without having to intervene.

A **Forrester** Consulting study commissioned by **Adobe** found that companies using applications of **Adobe Experience** Platform, such as Real-time CDP, Journey Optimiser, and Customer Journey Analytics, realised a 431% return on investment with a payback period of fewer than six months. It's a stark reminder of just how much AI-led analytics could influence marketing effectiveness. https://business.adobe.com/blog/the-latest/forrester-consulting-study-adobe-experience-platform-applications-deliver-a-431-percent-roi

Key B2B Marketing Metrics and What Artificial Intelligence Brings to the Mix

Lead Generation and Qualification

B2B marketing success begins with high-quality leads. AI improves lead generation by determining high-intent prospects and forecasting their likelihood to convert.

- Marketing Qualified Leads (MQLs) AI computes engagement data to predict which leads indicate buying intent, thus minimising the need for manual scoring.
- Sales Qualified Leads (SQLs): AI-based lead scoring models augment the process of identifying SQLs by analysing history, behaviour

and firmographics.

- Cost Per Lead (CPL): AI tools help you maximise your marketing spend by diverting budgets to well-performing channels.

Tools like **HubSpot AI** and **Marketo Engage** use machine learning to sift through mountains of data for signs of lead readiness and engagement.

Engagement & Conversion Data

Relating engagement behaviour from multiple touchpoints is essential in order to gain an understanding of how prospects consume your content. With the right AI-driven analytics, marketers can improve messaging, channels and timing to boost conversion rates.

- Click-Through Rate (CTR): A/B Testing powered by AI optimises subject lines, ad copy and content recommendations to maximise CTR.
- Time on Page and Bounce Rate: AI tools like **Google Analytics 4** can process how people are interacting with your content and recommend content optimisations to drive more engagement.
- Conversion Rate Optimisation (CRO): Tools like AI increase website conversion rates through personalisation of website experiences according to information about the visitor.

Revenue and ROI Metrics

When marketing works with AI, it helps us track which marketing activities drive sales and streamlining budgets.

- Customer Acquisition Cost (CAC): AI forecasts which channels offer the lowest CAC and reallocates spend based on that prediction.
- Customer Lifetime Value (CLV): Using purchase behaviour and engagement trends, AI models forecast potential CLV, enabling companies to concentrate on their high-value customers.
- Marketing ROI: AI-powered analytics solutions like **Salesforce** Marketing Cloud Intelligence enable real-time ROI assessments for campaigns.

Predictive Analytics and Insight By AI

Its ability to forecast future trends from past data is one of the biggest advantages of AI in marketing measurement. AI enables:

- Churn Prediction: AI can recognise symptoms of potential customer churn, helping companies develop proactive retention strategies.
- Sales Forecasting: AI platforms like **Clari** use data from the sales pipeline to forecast future revenues and improve marketing campaigns.
- Intent Data Analysis: Artificial Intelligence based solutions like **6sense** track buyer intent signals that allow marketers to reach out to prospects at the opportune time.

AI is transforming the way B2B marketers measure performance and tune strategies. By using AI-powered analytics, companies can better determine the quality of leads, the effectiveness of engagement and the impact of revenue.

What is an Attribution model in B2B marketing

One of the toughest elements of B2B marketing measurement is knowing which channels, messages, and touchpoints drive the most conversions. B2B purchasing journeys can be lengthy, with multiple decision-makers requiring dozens of interactions across multiple channels. Traditional attribution models, such as first touch or the last touch, simply fail to give a comprehensive look at how marketing influences results.

Artificial intelligence is revolutionising the way attribution modelling works as it can track intricate buyer journeys and attribute value to each interaction depending on its specific influence. Attribution (that is AI-driven) empowers marketers to see not only that something took place but also what, why, and how it affected that result.

Traditional Attribution Models and Their Limitations

There are several common examples of attribution models:

- First-touch attribution: Where all the credit goes to the first interaction (a paid search click or a website visit).
- Last-touch attribution: Success is attributed to the last action before purchase or conversion (such as when a product demo or an email message is clicked).

- Linear attribution: All touchpoints share credit equally.
- Time-decay attribution: Credit for the conversion would be distributed across all touchpoints, with more weight given to those closer to the time of conversion.

Although these models offer a sense of direction, they do not reflect variations in the influence of the channel, depth of engagement, or timing. These confines are especially troubling in B2B where deal cycles can last months or even years.

AI-powered Multi-Touch Attribution

Thanks to artificial intelligence, businesses can execute better and more dynamic attribution through multi-touch models. AI analyses all the points at which the customer has touched the marketing and sales ecosystem to determine which channels and messages have been the most fruitful in moving the lead down the line.

- Share-of-interaction attribution: AI intelligently measures the impact of each touch on an engagement or interaction to understand conversion rates and historical trends.
- Custom weight modelling: Through platforms such as AI-based **Google** Attribution and **Adobe**'s Attribution IQ, businesses can create their own custom attribution models, one that is aligned to the sales cycle and buyer journey.
- Channel performance forecast: AI forecasts what future conversion impact reallocating budget to channel will have to drive proactive optimisation.

Platform platforms like **Rockerbox**, **Dreamdata** and **Wicked Reports** facilitate AI-powered multi-touch attribution so that marketers can confidently visualise the true performance of cross-channel marketing.

Account Based Attribution in B2B

For those conducting account-based marketing (ABM & ABX), attribution is a little more complicated. AI makes ABM attribution easier by consolidating and analysing touch points at the account level as opposed to the individual lead level.

- Artificial intelligence-powered tools monitor which contacts at a target account have engaged with certain content, channels or campaigns.
- Influence is gauged at the department level, which accounts for the collective path to purchase.
- Sales and marketing get better aligned as both are afforded visibility to the most impactful touchpoints.

Platforms such as **Bizible** (now part of **Adobe**) and **Terminus** deliver sophisticated ABM attribution using AI that helps marketers optimise campaigns for high-value accounts.

Benefits of AI-Driven Attribution

By embracing AI-driven attribution models, businesses can:

- Find out which campaigns, channels, and assets are giving you the best ROI
- Allocate budget effectively across underperforming or over performing channels
- To make marketing investments more clear and accountable
- Uncover hidden touchpoints that move the needle most on conversions

AI-Powered Continuous Improvement Strategies

In today's high-velocity, data-rich B2B marketing environment, performance optimisation is not a post-campaign, one-time effort. It has evolved into a never-ending cycle fuelled by real-time analytics and automated optimisations. AI is one of the chief drivers, making it possible for marketers to track, measure, and optimise campaign performance with such speed and accuracy. AI enables companies to create feedback loops, pinpoint lagging elements, and test new strategies in an agile and scalable manner.

The types of AI-led continuous improvement strategies are changing the way we think about marketing experimentation, responsiveness and evolution.

Real-Time Performance Monitoring and Automated Adjustments

Historically, additional time was required to conduct fixed campaign duration and interval reviews in order to inform marketing optimisation. AI has outmoded this static model with live performance dashboards and automatic alerts when there are anomalies, underperformance or changes in buyer behaviour.

Platforms, such as **DataRobot** and **Anodot**, leverage AI to identify early symptoms of change in KPIs so that marketers can respond in time before problems become too daunting or opportunities are lost. These tools interpret historical baselines, predict with models and serve out new insights that would be drowned out in large data sets.

This enables marketing to:

- Move spending to more efficient channels
- Optimise underperforming messaging or creatives during campaign flight
- Act quickly in the face of new developments or a change in market, parties or competitors

A/B and Multivariate AI-Powered Testing

Testing is key to finding what works with a desired audience. AI doesn't just aid A/B and multivariate, simultaneously running tests. Instead, it automates test creation, real-time result analysis and variable adjustments based on results that are generated.

- Tools like **Optimisely** leverage machine learning to deploy the best content variations automatically, enabling the marketing team to scale the experimentation process with no human resources.
- Platforms like **Dynamic Yield** use responsive testing to automatically test headlines, images, calls to action and layers of personalisation, then serve up content according to visitor behaviour and segment-specific patterns.

Unlike manual testing, which relies on human intervention to understand and act on findings, AI-based testing is dynamic and self-optimising.

Predictive Content & Campaign Optimisation

AI helps marketers predict which content formats, topics, and channels will perform best on the basis of audience preference and historical performance. These estimates are continually updated as additional data is processed.

- **PathFactory** applies AI technology to recommend content that speeds buyer journeys based on real-time engagement signals and previous behaviours.
- Uberflip's AI-powered content recommendations engine personalises content, making sure content is presented in the right sequence and format.

This level of optimisation drives higher engagement, quicker funnel acceleration and better marketing ROI.

Closed-Loop Learning Systems

AI brings even more into the fold with the notion of closed-loop marketing. These systems continually capture information from customer interchanges, apply machine learning to discern patterns and feed those insights back into campaign planning and execution.

The closed-loop AI marketing system generally comprises:

- Data ingestion from CRM, web analytics, and ad platforms
- AI models for pattern recognition and trend analysis
- Predictive modelling and content recommendations
- Automated execution of optimised campaigns

The more these systems are used, the more accurate and efficient they become, which reduces waste and increases marketing precision.

Enablement of the Team And Automation of the Process

Aside from campaign optimisation, AI empowers team productivity by turning mundane tasks into automated actions, ranking priority tasks, and providing insight into where the next-best marketing opportunity resides.

For example:

- **Zapier AI** and **HubSpot**'s Operations Hub automate all your marketing flows, such as updating CRM fields, sending nurture emails, and triggering sales alerts based on lead scores.

- AI-enabled project management tools such as Asana Intelligence indicate due dates, highlight bottlenecks and support resource planning.

These functions liberate marketing managers to create strategies and develop creativity, leaving AI to do the heavy lifting of data processing and campaign optimisation.

Artificial Intelligence enables B2B Marketers to bring continuous improvement by opening up possibilities of real-time optimisation, scalable experimentation, and self-learning systems. This helps to see inefficiencies and increase the impact of the campaign, which allows your marketing to follow the behaviour of the audience and the dynamics of the market.

Agile Marketing & Performance Optimisation

With the increasing pace of change and reliance on data in marketing, new ways of working will replace long-cycle historical campaign planning with agile ways of working. Like its agile software development counterpart, agile marketing is focused on iterative development, active cross-functional teaming, very rapid experimental building, and being able to change. When combined with artificial intelligence, agile marketing is not only faster and more responsive but also vastly more informed.

AI enables teams to work in real-time, acting on data as it is produced, minimising 'gut feelings' and empowering marketers to constantly test, learn and scale what works. This fosters a culture of high-speed optimisation, with every campaign being a chance to learn and improve.

Principles of Agile Marketing

Agile marketing processes are founded on the following principles:

Sprints and iteration - campaigns are executed in quick order and can be adjusted and improved mid-cycle.

- Customer first: Ideas are led by ongoing customer feedback and behaviour.

- Data-informed decision-making: Performance data is monitored in real-time to guide decisions.
- Cross-functional collaboration: Marketing, sales, analytics, and product teams work together seamlessly to eliminate bottlenecks and align around shared goals.
- Test-and-learn culture: moving quickly to test hypotheses and implement results into future campaigns.

When you combine them and scale them with AI, these principles scale exponentially, stripping friction from the decision-making process and empowering teams to act quickly when performance, customer engagement, or market conditions change.

How AI Powers Agile Execution

Agile marketing is augmented in this environment by AI taking care of the data, extracting insights, and even making some decisions for you. This speeds up the feedback cycle and makes agile practices more enduring.

- AI will automatically optimise around the highest impact tasks based on resource capacity and business goals.
- Predictive models direct sprint planning by predicting the results of offered activities.
- Automation tools further allow marketers to test ideas without having to wait for full campaign deployments.

Resources such as Workfront by **Adobe** or **Monday.com** AI help manage agile workflows and incorporate AI recommendations into campaign planning and execution.

From Campaigns to Continuous Delivery

AI also allows marketers to shift away from traditional campaign thinking to continuous delivery. Instead of 'big bang' campaigns running for fixed periods, businesses can now operate always-on programmes that are iteratively optimised, with the performance tuned on the basis of live performance data.

This is made possible by:

- Real-time response monitoring: Tools like **Braze** and **Iterable** monitor how customers are responding and update campaigns in real-time.

- Personalisation engines: AI can help disciplines like **Dynamic Yield** or **Evergage** (now **Salesforce** Interaction Studio) to automatically serve content and optimise timing on a 1:1 scale.
- Predictive journey orchestration: AI maps optimal customer journeys and delivers content based on real-time behavioural inputs.

Culture Change: Creating an Agile, AI-Ready Team

It's the mindset shift, not the tools, that characterise agile marketing. AI assists in breaking down internal silos by offering shared visibility into data and by helping cross-functional teams agree on goals and outcomes.

Key enablers include:

- AI dashboards and collaboration platforms offer a single source of truth across teams
- Clear performance goals that connect individual KPIs to larger marketing objectives
- Training for data literacy to ensure teams can accurately interpret AI insights

Enterprising businesses from **Unilever** to **IBM** have been adopting agile marketing methods, which are powerfully enabled by AI, resulting in faster campaign delivery, improved marketing team collaboration and greater ROI on marketing spend.

Unilever deploys AI in its marketing to improve efficiency and agility. They produce real-time 3D digital twins of products that are accurate based on machine learning products using AI enabled technology, meaning they can be quickly adapted across channels and dramatically cut production costs and time. This process has enabled **Unilever** to produce content at half the speed and half the cost, showing the effect of AI technology on its adaptability in marketing. https://www.marketingdive.com/news/unilever-ai-marketing-bets-halve-production-costs-double-speed/742913/

Equally, **IBM** has adopted AI-driven agile marketing methods that have led to better decision-making and greater effectiveness around campaigns. **IBM**'s AI marketing solutions provide data to marketers in near real-time, which allows them to pivot strategies quickly and increase ROI. Such AI solutions aid in recognising actionable insight, optimising advertising exposure and personalising content for higher productivity and more effective marketing activities. https://www.ibm.com/think/topics/ai-in-marketing

By combining AI with agile marketing, the likes of **Unilever** and **IBM** have developed marketing tactics that are more adaptive, market-led, and data-smart, enabling them to react rapidly to changing market forces and consumer demands.

Agile marketing and AI work hand in hand. AI allows teams to adapt agile principles more easily and with greater confidence, just as agile practices liberate organisations to take full advantage of AI tools by promoting responsiveness, iteration, and experimentation.

The Future of AI in B2B Marketing Performance Measurement

As AI increasingly becomes part of the marketing landscape, the role of AI in performance measurement will evolve from a support to a strategic one. And if AI is contributing better campaign analysis and optimisation today, the near future will see it play the role of being the decision-making engine of choice, providing guidance that not only helps marketers assess what has worked but what should come next. This journey toward prescriptive analytics will change how performance is measured, optimised and applied in real-time.

The marketers who succeed in this environment are those who not only accept AI as a tool but also integrate it into their decision-making process.

Predictive and Prescriptive Evaluations of Performance

B2B marketers already benefit from predictive analytics, where they expect outcomes based on behaviour history. The next leap is prescriptive analytics, in which AI not only predicts what will happen but also tells you exactly what to do about it.

The likes of **Salesforce Einstein** or **Microsoft Dynamics 365** AI are leading the way in embedding predictive intelligence into CRM and marketing systems, which is supporting businesses to optimise touchpoints before they are even deployed.

This shift will enable marketers to:

- Test campaign results before going live with virtual testing environments
- Get AI recommendations on where to reallocate budgets, tweak targeting or messaging on the fly mid-campaign
- Forecast how buyers will behave in response to market signals, competitor activity, or economic conditions

Intelligent Marketing Decision Support Systems

Autonomous marketing systems are the future. These applications will reach high-scale, low-risk decisions without human involvement using AI models. Audience targeting, creative selection, and bidding adjustments will be automatised as standard optimisation measures.

- AI-driven media buying platforms such as The Trade Desk and **Adobe** Advertising Cloud are already utilising real-time bidding algorithms to enhance optimal ad placements at scale.
- The marketing automation platform of the future will apply reinforcement learning to learn continuously from campaign feedback loops, consumer responses, and channel performance.

Autonomous marketing will enable teams to operate at a higher level, where strategies and creativity are invented, and the ethical frameworks and rules of operation are policed.

Unified Data Ecosystems & AI Integration

Another defining development has been the shift toward unified marketing data ecosystems, where all the performance data from every channel and touchpoint is aggregated, cleaned, and processed in real-time for AI analysis.

Data foundations include platforms such as **Snowflake** and **Google**

BigQuery while AI tools such as **Looker Studio** and **Tableau** AI provide actionable insights across departments.

Marketers will work from an integrated source of truth for performance, enabling cross-functional teams to continue to act from a common set of insights.

This will eliminate data silos and minimise the reporting lag time, as well as allow for better synergy between marketing, sales, product and finance teams.

Ethical AI and Transparent Measurement Practices

As AI's efficiency measurement of performance becomes more powerful, transparency and ethics will be considered must-haves. Marketers should be vigilant that AI-based recommendations and optimisations do not develop a bias or cause unethical user behaviour and erode trust.

Best practices in ethical performance measurement include:

- Regularly auditing AI algorithms that are responsible for bias and transparency
- Human-in-the-loop for crucial decisions
- Empowering users to determine the extent to which their data is used to power AI-personalisation

The Marketer's Role in the AI-Driven Future

As AI takes over the technical analysis and decision-making, the human marketer's role will be focused on:

- Defining ethical limits and strategic orientation
- Interpreting AI findings in relation to business/story context
- Fostering innovation in creativity, brand and customer experience

The most effective marketing departments will be those that look at AI as a co-worker, to complement human skills, not replace them. They will marry the nuance of data to the authority of narrative to create durable, quantifiable relationships with an audience.

The future of marketing measurement is intelligent optimisation, which is predictive, prescriptive, and autonomous. The B2B marketers who forge the infrastructure, skills, and ethical guardrails for AI integration today will be the ones in the driver's seat tomorrow. With AI maturing, its role in performance calibration will shift from that of insight provider to strategic conductor, ensuring every move is made at the right time and with the right information and is in line with the long-term prosperity of the business.

Summary:
It's no longer the case that you measure and optimise marketing at the end of the quarter; now, you must do it in real-time. As AI continues to play a growing role in B2B marketing, it allows companies to do more than report on what happened; AI can also show why something happened and what to do as a result in a fraction of the time required to do it manually.

- Define performance indicators that reflect the business side and the customer behaviour side.
- Utilise AI-driven analytics to track performance in real-time and automate optimisation.
- Leverage multi-touch and account-based attribution models for a true understanding of how the business is being impacted.
- Use agile marketing to constantly test, learn, and iterate.
- Get ready for a world where AI offers predictive and prescriptive insights to inform strategy.

Marketers that embrace AI head-on and approach performance measurement differently have the opportunity to be best positioned to create value, demonstrate ROI and adapt at the speed of business.

Example in Action: Schneider Electric's AI-Powered Marketing Performance Strategy

Schneider Electric, a global leader in energy management and automation, has adopted an AI-powered approach to marketing performance measurement. The company uses a unified marketing automation platform that integrates AI to track engagement, optimise campaigns, and attribute value across channels and touchpoints.

By implementing predictive analytics and multi-touch attribution models, Schneider Electric has improved marketing alignment with sales and enhanced ROI visibility across global regions. Their agile marketing teams use AI insights to iterate in real-time, allowing campaigns to evolve continuously based on performance and customer feedback.

This data-centric approach has empowered Schneider Electric to enhance customer engagement, lower acquisition costs, and globalise marketing impact. https://www.salesforce.com/blog/ai-at-scale/

19. Where AI is Headed in B2B Marketing

"The future is already here, it's just not evenly distributed."

William Gibson, Science Fiction Novelist

AI in B2B Marketing – The Next Phase

The current wave of AI in B2B marketing has revolutionised how we identify intent, personalise journey stages, and automate content creation at scale. But this is just the start. What's coming is nothing short of a seismic shift from AI as a toolkit to AI as a strategic co-pilot and, ultimately, an autonomous orchestrator of business decisions.

Here are the five directions that I believe are defining the future of AI and B2B marketing:

1. From Keywords to Intent Graphs

Search, as we know it, is changing. Training search engines to understand queries contextually, not just semantically, is the work of large language models (LLMs). Instead of just analysing keywords, AI will map intent graphs, mental multidimensional constructs that will tell us what buyers want, even before the buyers know it themselves.

For marketers, it's the dawn of zero-click discovery, voice-activated questions, and AI-generated answer summaries in place of web pages. Content will have to be well organised, question-led, and ultimately highly value-focused in order to rank and resonate.

2. Generative Personalisation at Scale

We are shifting from segment-based personalisation to moment-based orchestration. Real-time behavioural and environmental signals will be fed into AI to personalise every touchpoint, the copy in an email, a landing page, a product demo, and a pricing offer, all at the moment, for the individual.

The likes of **Adobe Sensei**, **Salesforce Einstein** and **Meta Lattice** are early indicators of what's to come. Eventually, each B2B buyer will go through an experience that appears to have been custom crafted just for them, though in reality, it was formed by machines all the way through its creation.

3. Autonomous Agents and Always-On Marketing

One of the biggest upcoming changes is the rise of AI agents. Self-directing digital entities that can execute campaigns, run tests, handle customer queries and optimise content, all without any human intervention.

Already, the building blocks are being put in place by tools such as **AutoGPT**, **Cognosys** and **Salesforce**'s Prompt Studio. In the not too distant future, these agents will behave much like junior marketers - consuming strategy, understanding objectives and autonomously making decisions, all the while being under human oversight.

Marketing teams will go from content creators to being 'experience' architects and AI trainers to ensure agents act in brand-safe, customer-centric ways.

4. Emotion Recognition and Adaptive Engagement

As multimodal AI continues to gain popularity, you'll see more people using tools that analyse facial expressions, vocal tone and sentiment during video calls, webinars, or feedback loops. These inputs will help shape not only what message to send but also how and when to send it.

Picture AI-enabled sales enablement that adapts messaging according to the prospect's mood. Or webinars that drive various follow-up journeys based on real-time engagement at that moment. The algorithm itself will be hard-wired with empathy.

5. The Emergence of AI-Centric Operating Models

And finally, AI is not just a tool for execution, it is transforming the structure of marketing organisations. We will see the rise of:

- AI marketing strategists: mentors who directionally lead AI.
- Prompt engineers: experts who teach LLMs to produce on-brand, high-quality content.
- Ethical AI committees: teams that ensure proper compliance, fairness, and explainability in the use of marketing data.

Businesses that re-engineer their operating model to fully embrace AI while retaining a focus on human intelligence and judgement will be more agile, more productive and more competitive.

Navigating What's Next

Marketing isn't going away. It will evolve. The best-case scenario for AI in B2B is one of human innovation and collaboration with machine intelligence. The humans will still own strategy. However, the delivery, optimisation, and scale will increasingly be left to AI systems that never rest.

This transition will not occur overnight. But it will arrive more quickly than many expect. The marketers who emerge victorious in this new landscape will be those who don't simply add AI to the mix but who instead re-imagine what marketing can be because of it. What will your role be in determining what comes next?

The future is not just AI-driven, it's AI-partnered. And it's coming fast.

20. Blazing the Trail for B2B Marketing in the Future

"The best way to predict the future is to invent it."

Alan Kay, Computer Scientist

B2B marketing today is not just about creativity and communication; it is a scientific pursuit where technology, data and a relentless commitment to innovation are equally important. When companies are dealing with rapidly accelerating AI, buyer expectations are shifting, and there is a need to be more sustainable. Never has it been more important for marketers to have a forward thinking, flexible approach.

This book has hopefully led you through the confusing and fun 'still new' landscape of modern B2B marketing. It has also been discussing the art of storytelling and how to succeed with brand narratives, the importance of tapping into the numbers, and the impact of innovation as it continues to lobby upon how to win in a world of global giants. There is an emphasis on the need to accept change, not as a threat but as an opportunity to redefine what success looks like in a growing field.

To thrive in this new era, marketing leaders need to evolve into multidimensional strategists. This isn't just about expanding skill sets, it's about embracing a new professional mandate. The best future-ready marketers will embody four critical roles:

- **Strategic Translator** – Bridge commercial objectives with creative outputs. Great marketers don't just execute strategy; they interpret it, frame it, and translate it into buyer value.
- **Tech Navigator** – Champion the responsible adoption of AI and automation. Not every tool should be used, and not every output

should go unchallenged. Navigating the landscape is as important as exploiting it.

- **Cultural Custodian** – Maintain message integrity and internal cohesion. Marketers are now brand stewards across continents, platforms and product teams; coherence is currency.
- **Data Synthesiser** – Move from measurement to meaning. Surface insights that matter and challenge teams to act on them, not just report them.

These roles aren't theoretical. They're what top-performing CMOs are already adopting as AI, complexity and stakeholder expectations reshape what it means to lead marketing.

It will take courage, innovation and cooperation in the road ahead. It requires marketers to marry the precision of data with the creativity of human insight, leveraging AI and machine learning technologies to refine personalisation, automate operations, and ultimately deliver unrivalled customer value. It also requires laser-sharp attention to sustainability, ethical behaviour, and social responsibility; marketing should help make the world a better place.

With Sam Altman and Jony Ive joining forces as OpenAi acquires io Products, the next generation of tools, devices and platforms are just around the corner. The AI space is going to get bigger, faster and more competitive.

As you ponder the ideas and insights in this and other works, be transformed, act boldly and lead with purpose. The future of B2B marketing is for those who don't just predict change but make it happen.

The coming challenges will produce winners and losers, no doubt. To become a winner, you need to think differently and welcome AI as your new colleague in life.

Essentially get yourself 'Rewired' in the new era of marketing.

21. Postscript: AI Governance and Responsible Innovation

Artificial intelligence is increasingly at the core of how B2B marketers work. From customer segmentation and content creation to predictive analytics and campaign orchestration, AI has shifted from the periphery of experimentation to the core of daily execution.

But with great power comes new responsibility. AI is not made in a vacuum. It learns from data, runs on algorithms and is used by human teams, each of which offers opportunities for bias, abuse and unintended consequences. As AI is becoming ever more autonomous and integrated, the issues around transparency, accountability, and ethics must be tackled with the same fervour that marketers use to address brand, compliance and precision of measurement.

This chapter isn't a warning, it's a call to leadership. The future of marketing will be shaped not only by how we deploy AI, but by how responsibly we govern it. Marketers don't need to be AI engineers. But they must become AI governors, capable of asking hard questions, documenting risk, and defending decisions.

Why AI Governance Matters

The risks of AI are real. These include:

- Algorithm Bias: If the models AI uses to make decisions are based on biased data, they can lead to unintentional discrimination against marginalised groups or people in certain demographics
- Lack of Transparency: Complicated models can transform into "black boxes" where it is difficult to explain why decisions are made
- Misinformation: Generative tools might generate incorrect or misleading content, particularly when used without human oversight.

- Privacy Breaches: Bad data understanding or model training might contribute to using private or non-consented data.
- Erosion of Trust: If customers or partners begin to think that automation has supplanted human integrity, image and brand will suffer.

AI doesn't displace the need for judgment, it multiplies it. B2B marketing leaders must develop strong governance frameworks that dictate ethical development, deployment, and oversight of AI systems.

Foundations of Responsible AI

Although AI governance is a challenging and growing field of study, there are a few basic principles that have become acceptable as a basis for the ethical deployment of AI. These include:

- Fairness: AI cannot make decisions that are based on race, gender, geography, or any protected characteristic
- Accountability: All AI-augmented processes must have human oversight
- Transparency: Organisations need to be able to define how & why AI reaches decisions
- Privacy: All the data used in AI models needs to be ethically sourced, permissioned and secured.
- Security: Systems need to be secured to avoid tampering, misuse or external attacks.

Standards like the OECD AI Principles https://www.oecd.org/going-digital/ai/principles/ and the NIST framework https://www.nist.gov/itl/ai-risk-management-framework provide a good template for many businesses to adopt internationally.

The Alan Turing Institute is also conducting research into Artificial Intelligence, its applications and implications for society. https://www.turing.ac.uk/research/research-programmes/artificial-intelligence

Creating a Model for Marketing AI Governance

Marketing teams should not shy away from stepping up if legal or IT fail to lead the way. A practical governance model for AI in B2B marketing might include:

- Use-case classification: Categorisation of AI systems by risk (e.g. low-risk automation vs. high-risk decision-making tools)
- Approval workflows: Needing to have AI-enabled content or outputs reviewed before external publication
- Model documentation: Maintaining records of how each AI model was trained, what data was used, and how its outputs are validated.
- Disclosure policies: Communicating with customers and stakeholders about when content is AI-generated or decisions are automated
- Training programs: Making marketers aware of what AI can and cannot do for them

Governance is not here to add friction but to protect the brand, ensure fairness and build trust.

Responsible AI Tools and Technology

There are now a number of tools to assist companies in auditing and managing AI ethically:

- Linux Foundation AI Fairness 360: An open source toolkit that can help detect and remove bias in machine learning models. https://ai-fairness-360.org/
- **Microsoft Fairlearn**: Tools for assessing fairness and strategies for mitigating unfairness within models. https://fairlearn.org/
- **Google**'s What-If Tool: Visualises how models respond to various inputs and detects unexpected behaviour. https://pair-code.github.io/what-if-tool/

They can be delivered to promotion workflows by the use of data or analytics groups. They help make sure that the AI systems marketers depend on are not just powerful but also responsible.

AI Regulation: What's Coming

Governments are starting to legislate AI. The EU AI Act and the UK's nascent regulatory framework will soon set legal expectations in relation to transparency, explainability and oversight. B2B marketers employing AI for personalisation, targeting, or automation will also be impacted head-on. Be ready, marketers should start prepping by:

- Auditing existing tools and processes through the lens of AI
- Establishing responsible data use policies
- Factoring AI into supplier and technology evaluations
- Reviewing privacy and consent frameworks based on AI data practices

Getting ahead of regulation is not only a legal matter; it is a matter of reputation.

Cultural Integration of Responsibility

Watching over AI shouldn't be an afterthought. It has to be part of the marketing culture, influencing the way decisions are made, tools are chosen, and results are interpreted.

Leaders can facilitate the embedding of responsible innovation by:

- Integrating AI ethics into onboarding and training
- The hosting of cross-functional forums on AI risks and opportunities
- Inspiring teams to test the outputs of AI and apply critically thinking
- Rewarding and recognising good practices, not just performance metrics

AI is not neutral. Its outputs are the outputs of its inputs, designers and use cases. Marketers have a responsibility to ensure that these systems reflect the values of their customers, brand and the society in which they operate.

AI will define the future of B2B marketing. But how it's formed, by whom, and by what rules, and for what purpose remains in our hands.

Responsible innovation is not a straitjacket. It's a competitive advantage. The organisations that manage AI responsibly and with care regarding the future, transparency, and fairness won't just be risk-avoidant; they'll be more trusted, achieve better outcomes, and lead the market.

Marketing has always been about relationships. AI does not change that. It only raises the stakes. This book began with a provocation, that B2B marketing must rewire its thinking. It ends with a responsibility: to build systems that earn trust in a world where machines now shape meaning.

22. The Rewired CMO Toolkit

"The last best experience that anyone has anywhere, becomes the minimum expectation for the experience they want everywhere."

Scott Brinker, VP Platform Ecosystem at HubSpot and Editor at chiefmartec.com

10 Things Every Modern B2B Marketing Leader Needs to Do Well

The role of the CMO has changed. It's no longer just about campaigns and leads. It's about building trust, leading teams through change, and making smart, long-term bets in a fast-moving world.

Here are 10 things today's B2B marketing leader needs to have in their toolkit:

1. **Understand how people really make decisions.**
   Get beyond the funnel. Learn the psychology behind trust, emotion, and behaviour and use that to guide your message.

2. **AI as a co-pilot, not a black box.**
   Use AI to help you think better, spot trends, predict needs, and test faster, but always stay in the driver's seat.

3. **Treat the brand like a strategic asset.**
   Build brand equity with the same discipline as your pipeline. The brand is what keeps you in the room when pricing isn't.

4. **Make content that's genuinely useful.**
   Content isn't filler. It's how you show up, teach, support, and earn trust before anyone talks to sales.

5. **Align with revenue, not just leads.**
   Marketing doesn't stop at the MQL. Partner with sales, customer success, and finance to drive impact that lasts.

6. **Think globally, act locally.**
   Your message needs to travel but it also needs to feel right for each market. AI can help localise at scale, but empathy has to guide it.

7. **Take smart creative risks.**
   Safe rarely stands out. Try new formats. Break patterns. Use data to learn, not to play it safe.

8. **Keep ethics front and centre.**
   Whether it's how you use data or how you train your AI tools, your customers need to know they can trust you.

9. **Orchestrate across the whole journey.**
   Don't just market at the top of the funnel. Shape how your audience feels all the way from awareness to loyalty.

10. **Stay curious and help your team do the same.**
    Keep learning. Build a culture where experimentation is normal and growth is shared.

Final Word:
Being a modern CMO isn't about having all the answers; it's about asking better questions, staying human in a world of machines, and building a brand that earns its place in the future.

Glossary of Key Terms, Tools and Concepts

AI & Data Intelligence Platforms

6sense – Predicts which B2B accounts are in-market using AI and intent signals.
AI Fairness 360 (IBM) – An open-source toolkit for detecting and mitigating bias in machine learning models.
AI Ops Layer – An operational AI layer that manages automated systems, often used in Martech stacks.
Bombora – Tracks content consumption across a shared network to reveal buying intent.
Cognism – A B2B data platform offering sales intelligence and lead generation through AI-powered tools.
Gainsight – Platform for customer success teams to improve onboarding, monitor account health, and reduce churn.
Google AI - A division of Google dedicated to artificial intelligence.
Heap – Automatically collects and visualises user behaviour on digital products.
Hotjar – Tool for understanding how users interact with websites through heatmaps and recordings.
Infer – Uses machine learning to assess lead quality and conversion probability.
Segment (Twilio) – Customer Data Platform that centralises and connects customer data from all sources.
Totango – Tracks customer engagement and adoption across their life cycle.

Marketing Automation, CRM & ABM Platforms

ActiveCampaign – Automation platform blending email, CRM, and predictive content suggestions.
Adobe – A global software company that produces digital media and marketing tools.
Adobe Sensei – AI framework enhancing creativity and personalisation across Adobe products.
Autopilot – Visual customer journey builder for cross-channel automation.

Demandbase – An account-based marketing platform offering real-time intent data and personalisation.
HubSpot – User-friendly CRM and marketing platform for automation and lead nurturing.
Marketo Engage (Adobe) – Marketing automation platform for campaign management and lead nurturing.
Pardot (Salesforce) – B2B marketing automation platform for lead scoring and ABM.
Salesforce / Salesforce+ / Salesforce Einstein – CRM leader with AI tools for forecasting and personalisation.
Terminus – ABM platform for multichannel engagement of target accounts.

Content Creation & Creative AI

Canva – Design tool with AI for generating visuals and templates.
Jasper – AI writing assistant for branded content creation.
Persado – AI platform optimising marketing language for emotional impact.

Customer Experience & Journey Orchestration

Dynamic Yield – Personalisation platform using real-time user data.
Insider – AI-powered journey orchestration platform for multichannel campaigns.

Data Visualisation & Analytics

Clari – Revenue intelligence platform for pipeline visibility and forecasting.
Google Analytics 4 – Tracks user behaviour with built-in machine learning.
Looker Studio (Google) – Customisable data visualisation platform.
Power BI (Microsoft) – BI tool for visualising and sharing insights.
Tableau (Salesforce) – Data visualisation tool used for dashboards and business insights.

Customer Data Platforms (CDPs)

BlueConic – CDP that unifies profiles and predicts behaviour.
Lytics – ML-powered CDP for segmentation and personalisation.

Integration & Workflow Tools

RudderStack – Syncs customer data across tools and teams.
Talend – Integrates and governs business data.
Tray.io – Low-code platform for integrating data and systems.
Workato – Enterprise automation platform for workflows.
Zapier – A no-code platform for connecting apps and automating tasks.

Social & Community Insights

Brandwatch – Social listening platform for sentiment and trends.
Lexalytics – NLP tool for extracting insights from text.
LinkedIn / LinkedIn B2B Institute – Professional network and research hub for B2B marketing.
Sprinklr – Customer experience platform for social and review management.

Conversational Intelligence

Gong – Sales conversation analysis tool for coaching and insights.
Outreach – Sales engagement platform combining AI and automation.
Salesloft (Drift) – Sales platform offering engagement and conversation insights.

Strategy, ESG & Trust Frameworks

CLG (Community-Led Growth) – Go-to-market model leveraging engaged communities.
Ethical OS – Toolkit to anticipate tech ethics risks.
Fairlearn (Microsoft) – Open-source tool to evaluate and improve AI fairness.
Google What-If Tool – Interface for ML model testing and bias detection.
Growth Guilds – Internal learning networks to scale innovation.
OECD AI Principles – Guidelines for responsible AI use.

Companies Mentioned in Case Studies or Strategic Context

Bynder – Digital asset management platform for brand consistency.
Capgemini – Global tech consultancy cited for sustainability research.
Cisco – Enterprise technology leader in communications.
Corporate Ink – B2B Tech PR and Marketing Agency.
Dreamdata – B2B Activation & Attribution Platform.

EcoVadis – Procurement platform for ESG ratings.
Grace & Stella – Beauty brand featured in affiliate marketing case study.
IBM – Enterprise tech firm offering AI and ethics tools.
Intuit Mailchimp – Email marketing platform owned by Intuit.
Microsoft – Tech company providing AI, cloud, and governance tools.
Monday.com – Work OS for planning and collaboration.
Piwik Pro – Privacy aware analytics tool.
Plausible – Privacy aware analytics alternative to Google analytics.
Seismic – Enterprise platform for sales content automation.
Sedex – Ethical compliance platform for global supply chains.
Siemens – Engineering firm embedding ESG in product strategy.
Unilever – Consumer goods brand with strong B2B sustainability positioning.
Versa Marketing – Affiliate marketing agency in B2B case study.

Core Frameworks & Mental Models

Throughout the book, several key models help reframe the role of marketing in today's AI-enabled B2B world. These are not just frameworks, they're ways of seeing the market, the buyer, and your team differently.

1. The Trust Pyramid

Trust is earned in layers: relevance, consistency, empathy, and proof. Each level builds emotional confidence in your brand, especially in high-risk B2B decisions.

2. The 95:5 Rule (LinkedIn B2B Institute)

Only 5% of your audience is in-market now. The rest are future buyers. Build mental availability so your brand is remembered when they're ready.

3. Gut-to-Data Confidence Curve

Great marketing uses both instinct and evidence. The curve maps how teams manage gut feelings and data-informed evidence - without losing creativity.

4. The Invisible Funnel

Most buying signals now happen before you see them: peer chats, dark social, anonymous research. The traditional funnel is still there - but it's mostly hidden.

5. Flywheel Thinking

From campaigns to systems. Growth today isn't linear - it loops through customer experience, advocacy, and feedback to create momentum.

6. Brand as Strategic Infrastructure

Brand is no longer just communication, it's a layer of trust, memory, and differentiation that supports every team and channel.

7. Experience-Led Growth

Buyers don't remember every message, they remember how they felt. Your marketing is only as effective as the experience it creates.

8. The Augmented Marketing Leader

The new CMO isn't replaced by AI, they're elevated by it. Leading modern marketing means knowing when to trust machines and when to lean on human judgment.

Source Guide

This book references a wealth of insights from industry publications, thought leaders, and proprietary research. Key sources include:

Academic & Strategic Thinkers

David Aaker – The Future of Purpose-Driven Branding – Key voice on brand purpose and long-term brand value.
Sam Altman - CEO of OpenAi.
Jay Baer – Customer Experience Expert.
Les Binet & Peter Field – The Long and the Short of It – Leading voices on balancing long- and short-term marketing effectiveness.
Scott Brinker – VP Platform Ecosystem at HubSpot and Editor at chiefmartec.com.
Warren Buffett – An American Investor and Philanthropist.
Sarah Carter & Les Binet – How Not to Plan – Practical guide on strategic missteps to avoid in marketing planning.
W. Edwards Deming – The Father of Total Quality Management.
Peter F Drucker – Grandfather of Modern Marketing.
William Gibson – Science Fiction Novelist.
Brian Halligan – Co-Founder of Hubspot.
Ann Handley – Digital Marketing Pioneer.
Alan Kay – Computer Scientist.
Avinash Kaushik – Digital Marketing Evangelist.
Grace Kite – Economist focused on ROI and econometrics in marketing decision-making.
Philip Kotler – Marketing 4.0 – A transition framework from traditional to digital and data-driven marketing.
A.G. Lafley & Roger L. Martin – Playing to Win – Defines what strategy is (and isn't) through a business and competitive lens.
Jon Miller – Co-Founder of Marketo and Engagio.
Michael Porter – Founder of the Modern Strategy Field.
Joe Pulizzi – Content Marketing Expert.
Dr. Mark Ritson – A modern marketing guru – Renowned for challenging fads and reinforcing marketing fundamentals with clarity and commercial rigour.

Richard Rumelt – Good Strategy Bad Strategy – Explains the discipline of coherent strategic thinking and execution.
Howard Schultz – CEO of Starbucks.
Prof. Byron Sharp – How Brands Grow – Foundational research on brand growth, mental availability, and distinctiveness in both B2C and B2B.
Simon Sinek – Leadership Expert.

Psychology, Behaviour & Persuasion

Robert Cialdini – Influence: The Psychology of Persuasion (now with Unity as the 7th principle) – Explores human biases and heuristics relevant to B2B decision-making.
J. Jonathan Gabay – Practical Digital Marketing and AI Psychology – Connects cognitive psychology to AI-powered digital marketing.
Seth Godin – This is a Strategy and ongoing blog that focuses on reflections and observations that can help you approach problems, make decisions and create change.

Marketing Practitioners & Industry Voices

Prof. Scott Galloway – The Four – Examines platform dominance and tech economics. And No Mercy No Malice https://www.profgalloway.com/
David Meerman Scott – The New Rules of Marketing & PR – Advocate for inbound, attraction-based strategies and real-time relevance. https://www.davidmeermanscott.com
Jon Miller – Co-founder of Marketo and Engagio, instrumental in developing modern account-based marketing (ABM).
Dr. Mark Ritson - Marketing Week Columns. https://www.marketingweek.com/mark-ritson
Jes Scholtz – Operational Marketing Playbook – Offers a scalable, systematised approach to marketing operations and measurement.

Reports, Frameworks & Market Research

LinkedIn B2B Institute – Insights on the 95:5 rule, buyer behaviour, and long-term brand strategy.
Forrester Research – Forecasts and insights on B2B trends, trust, and revenue growth.

HubSpot – State of Inbound Marketing Report – Regular benchmarking of content, SEO, and automation trends.

McKinsey & Company – Research on AI, digitisation, and marketing productivity.

Gartner – B2B buyer journey analysis, tech adoption, and decision-making complexity.

Salesforce – State of Marketing Reports – Benchmarks on Martech, AI, and customer-centric strategy.

www.ingramcontent.com/pod-product-compliance
Lightning Source LLC
LaVergne TN
LVHW022158200926
848481LV00006B/1779